204

Topics in Current Chemistry

Springer-Verlag Berlin Heidelberg GmbH

Organosulfur Chemistry I

Volume Editor: Philip C. B. Page

With contributions by
B. Cid de la Plata, P. Metzner, J. L. G. Ruano

Springer

This series presents critical reviews of the present position and future trends in modern chemical research. It is addressed to all research and industrial chemists who wish to keep abreast of advances in the topics covered.

As a rule, contributions are specially commissioned. The editors and publishers will, however, always be pleased to receive suggestions and supplementary information. Papers are accepted for "Topics in Current Chemistry" in English.

In references Topics in Current Chemistry is abbreviated Top. Curr. Chem. and is cited as a journal.

Springer WWW home page: http://www.springer.de
Visit the TCC home page at http://link.springer.de/

ISSN 0340-1022
ISBN 978-3-662-14747-4 ISBN 978-3-540-48956-6 (eBook)
DOI 10.1007/978-3-540-48956-6

Library of Congress Catalog Card Number 74-644622

Originally published by Springer-Verlag Berlin Heidelberg New York in 1999
Softcover reprint of the hardcover 1st edition 1999

Cover design: Friedhelm Steinen-Broo, Barcelona; MEDIO, Berlin
Typesetting: Fotosatz-Service Köhler GmbH, 97084 Würzburg

SPIN: 10649335 02/3020 – 5 4 3 2 1 0 – Printed on acid-free paper

Volume Editor

Prof. Philip C. B. Page
Department of Chemistry
Loughborough University
LE11 3TU Leicestershire, UK
E-mail: P.C.B.Page@lboro.ac.uk

Editorial Board

Topics in Current Chemistry
Now Also Available Electronically

For all customers with a standing order for Topics in Current Chemistry we offer the electronic form via LINK free of charge. Please contact your librarian who can receive a password for free access to the full articles by registration at:

http://link.springer.de/series/tcc/reg_form.htm

If you do not have a standing order you can nevertheless browse through the table of contents of the volumes and the abstracts of each article at:

http://link.springer.de/series/tcc

There you will also find information about the

- Editorial Board
- Aims and Scope
- Instructions for Authors

Preface

Organosulfur Chemistry has enjoyed a renaissance of interest over the last several years, fuelled by its impact in the areas of heterocyclic and radical chemistry, and particularly stereocontrolled processes including asymmetric synthesis. One result of this resurgence of interest in the field is a rapidly escalating number of related publications. These volumes are intended to provide coverage of some of the highlights of contemporary organosulfur chemistry chosen from the entire range of current activity.

The first volume begins with a comprehensive review by Prof. José Luis Garcia Ruano and Dr. Belén Cid de la Plata of asymmetric cycloaddition mediated by sulfoxides, including dipolar and other processes in addition to Diels-Alder chemistry. It is followed by a discussion of the synthetic uses of thiocarbonyl compounds by Prof. Patrick Metzner.

Volume 2 begins with a thorough survey of sulfur radical cations, covering their synthesis, structure, stability, and reactivity, by Prof. Richard Glass. Prof. Naomichi Furukawa and Prof. Soichi Sato describe recent developments in the area of hypervalent organosulfur compounds, and the volume is completed by a discussion of the chemistry of thiophene 1,1-dioxides by Prof. Juzo Nakayama and Prof. Yoshiaki Sugihara.

Leicestershire, July 1999 Philip C. B. Page

Contents

Contents of Volume 205

Organosulfur Chemistry II

Volume Editor: Philip C. B. Page
ISBN 3-540-65729-0

Asymmetric [4+2] Cycloadditions Mediated by Sulfoxides

José L. García Ruano · Belén Cid de la Plata

Departamento de Química Orgánica (C-I), Universidad Autónoma de Madrid, Cantoblanco, E-28049-Madrid, Spain
E-mail: joseluis.garcia.ruano@uam.es

Asymmetric [4+2] cycloadditions involving the use of enantiomerically pure sulfoxides as the main controller of stereoselectivity are reviewed. Diels-Alder reactions with sulfinyldienes and sulfinyldienophiles, 1,3-dipolar reactions with vinylsulfoxides, and hetero-Diels-Alder reactions are the main objectives of this review. The influence of the sulfinyl group on the reactivity, regioselectivity, and especially stereoselectivity of these reactions has been mainly considered in order to understand the synthetic scope and limitations of the sulfinyl group acting as a chiral inductor in these reactions. Apparent deficiencies in proposed stereochemical models have been also highlighted.

Keywords. Asymmetric Diels-Alder; Sulfoxides; Asymmetric [4+2] cycloadditions; Sulfinyl dienophiles; Sulfinyldienes; 1,3-Dipolar reactions; Hetero Diels-Alder.

1 Introduction

During the last two decades, the asymmetric cycloaddition reaction has become one of the most powerful tools in asymmetric synthesis due to its capability of creating up to four chiral centers in one step in a highly stereoselective manner. The vast majority of the published work deals with the use of Diels-Alder reactions [1], but other different cycloaddition processes have also been investigated. The sulfinyl group had scarcely been a subject of interest in conventional Diels-Alder reactions due to its low ability as an activating group of dienophiles. Nevertheless, the intrinsic chirality and chemical versatility of the group, its configurational stability under quite varied conditions [2], and its ease of introduction into organic molecules in an optically pure form [3] drew the attention of many researchers working in different fields of asymmetric synthesis. In particular, the large steric and stereoelectronic differences between the substituents of the stereogenic sulfur atom suggested high potential for differentiation of the diastereotopic faces of neighboring double bonds, which prompted the investigation of the use of this functional group in asymmetric cycloadditions.

There are several reviews concerning the use of the sulfoxides in asymmetric synthesis. Many of them [4] deal with selected topics where cycloadditions are excluded or minimally considered due to the low importance of these reactions at the date of their publication (the first paper concerning an asymmetric Diels-Alder reaction of vinyl sulfoxides [5] was published in 1983). With respect to the most recent reviews, these are usually comprehensive and include sections devoted to cycloadditions (mainly Diels-Alder reactions). This is the case in the excellent and recent compilations of Carreño [6] (oriented toward the applications of sulfoxides to asymmetric synthesis of biologically active compounds) and of Solladié and Carreño [7], as well as that contained in the recent series "Stereoselective Synthesis" from Houben-Weyl [8]. There are also several reviews devoted only to cycloaddition reactions. This is the case in the survey by De Lucchi and Pasquato [9] of the role of different sulfur functions (including sulfinyl groups) in activating and directing olefins in cycloaddition reactions. This revision includes papers related to the use of racemic sulfoxides as well as those on optically pure substrates, up to 1987. In addition, Koizumi has published several reviews devoted exclusively to the synthesis and Diels-Alder reactions of vinyl sulfoxides [10]. Despite the fact that these reviews are mainly focused on the research of the authors, they are of great interest because Koizumi is one of the pioneers and is arguably the most important researcher in the use of optically pure sulfoxides in asymmetric cycloadditions. Finally, the preparation of optically pure sulfinyl dienes and their use in asymmetric synthesis has recently been reviewed by Aversa et al. [11].

Bearing in mind all the above information, this review is an attempt to present the results so far reported in asymmetric cycloadditions mediated by the sulfinyl group. This presentation is intended to emphasize those aspects related to the reactivity and stereoselectivity of these reactions starting from optically pure substrates. Results obtained from racemic sulfoxides are only explicitly considered in those cases allowing clarification of the behavior of the optically pure compounds, or to justify the absence of results in this field. Other aspects, such as the synthesis of alkenylsulfoxides or sulfinyldienes, are not specifically considered, because they are adequately covered in the comprehensive reviews on sulfoxides indicated above. In any case, most of the papers concerning cycloaddition reactions often contain direct or indirect information about the synthesis of the compounds used as the starting materials. Otherwise, the use of the adducts resulting from these cycloadditions as valuable synthetic intermediates, to prepare some natural products or interesting functional moieties, is specifically mentioned and fully referenced, but the complete reaction sequence connecting starting materials and final products is usually omitted.

We have also attempted to provide a vision of the scope and limitations of the sulfoxides in asymmetric cycloadditions. In this sense, we emphasize the main problems derived from its use indicating those features that cannot be explained with the currently accepted electronic description of the sulfinyl group. For this reason we have in some cases included comments on papers revealing discrepancies with the previously reported information, or have raised questions that have not been discussed by the authors in the original references. The final section includes a full discussion of the reactivity and stereoselectivity of the cycloadditions mediated by sulfoxides on the basis of the available information, including our own point of views about these issues.

This review is divided in five main sections. The first deals with the Diels-Alder reaction and is considerably wider than the others because most of the work has been carried out in this field. Sulfinyl ethylenes as dienophiles (Sect. 2.1) and sulfinyl dienes (Sect. 2.2) are the substrates considered. Sections 3 and 4 are devoted to the relatively small number of papers concerning the hetero-Diels-Alder processes (involving either heterodienes or heterodienophiles), and the use of α,β-unsaturated sulfoxides as dipolarophiles in 1,3-dipolar cycloadditions, respectively. The fourth section contains some relevant information concerning different asymmetric cycloadditions mediated by sulfinyl groups remote to the reaction center. In the last section (Sect. 6) we summarize the main problems remaining unsolved concerning mechanistic models and the advantages derived from the use of the sulfinyl group as chiral inductor in cycloaddition reactions.

In order to discuss the diastereoselectivity of these reactions, we use two different terms: *endo/exo*-selectivity and π-facial selectivity. The first is used to describe the mode of approach of the reagents (*endo* or *exo*) using as a reference the sulfinyl group. This criterion is maintained even in those cases that the sulfinylated substrate contains other functional groups having priority over sulfinyl, according to the sequence rules. The second term is used to describe the selectivity of the cycloaddition at each diastereotopic face of the enantiomerically pure sulfinylated fragment (diene, dienophile, or dipolarophile). According to the most widespread belief, the π-facial selectivity of these cycloadditions is mainly governed by steric factors. Therefore, the favored approach of the reagent will take place at the less hindered face of the sulfinylated substrate.

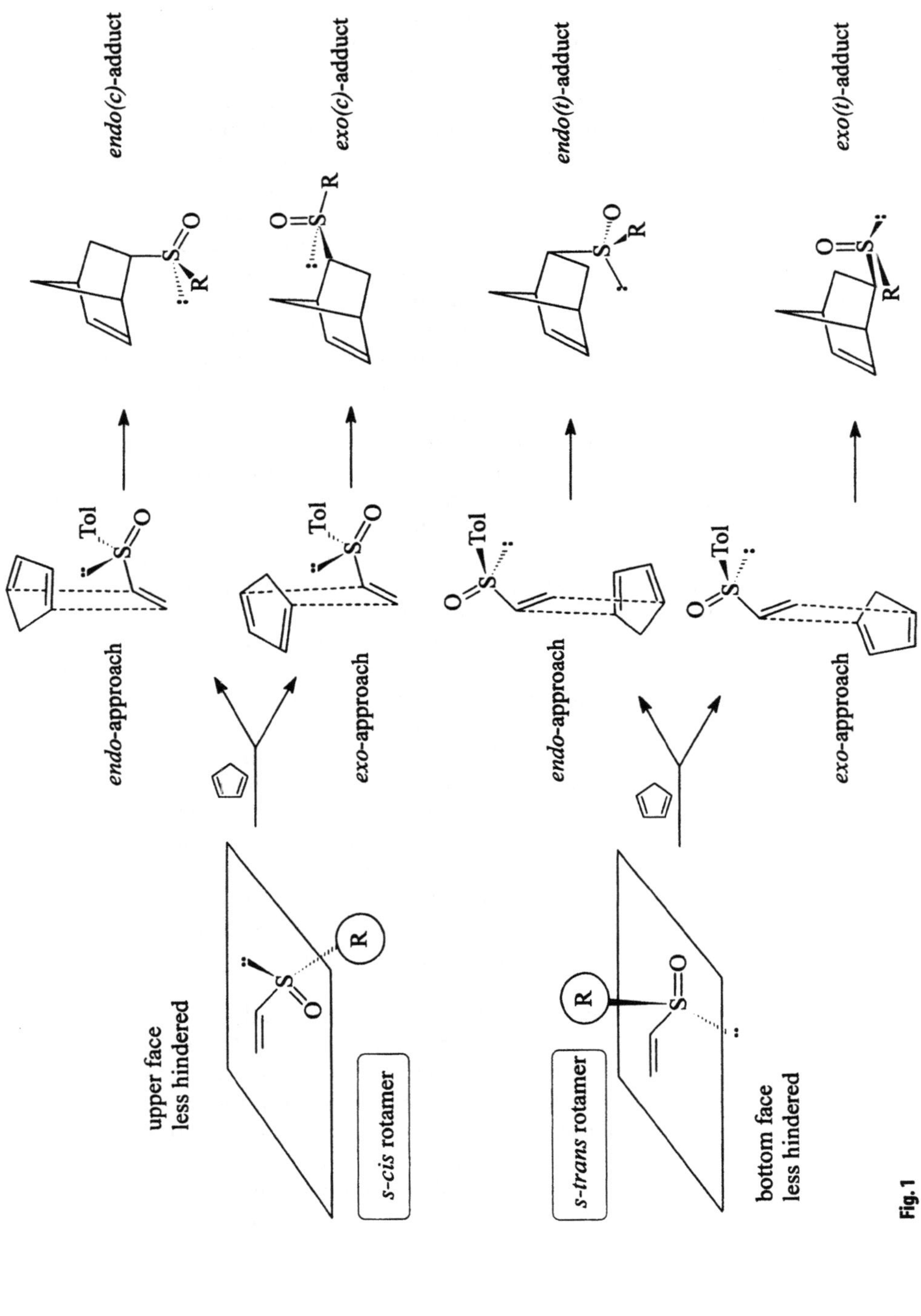

Fig. 1

In Fig. 1 it can be seen that any face of an alkenyl sulfoxide will be more or less hindered depending on the preferred conformation around the C-S bond. In this sense, the rotamers with the sulfinyl oxygen adopting *s-cis* or *s-trans* arrangements with respect to the double bond (Fig. 1) are used as the starting point of the majority of the stereochemical discussions. The rather large steric differences between diastereotopic faces of the double bond for both conformations (see Fig. 1) suggest that the approach of the non-sulfinylated reagent could take place only toward the less hindered one (that supporting the lone electron pair). Based on this assumption, we have added the letters (*c*) or (*t*) to the *endo* or *exo* prefixes to complete the stereochemical description of adducts. They denote the *s-cis* or *s-trans* conformation of the sulfinyl oxygen that must be considered to recognize the less hindered face of the sulfinylated reactant at which the *endo* or *exo* attack of the other reactant has taken place to form the designated adduct. Thus, for instance, the *endo*(*c*) term must be used to designate those adducts resulting from the *endo*-approach of dienes to the less hindered face of the sulfinyl dienophiles in an *s-cis* conformation, whilst the *exo*(*t*)-adducts derive from the sterically favored *exo*-approach of dienes to the *s-trans* rotamer of the dienophile. In those cases where conformations different to *s-cis* or *s-trans* must be considered to explain the experimental results, their evolution is assimilated into one of the two above designated, thus allowing retention of the nomenclature used. Finally we must remark that the situation depicted in Fig. 1 would be exactly the inverse for sulfoxides exhibiting the opposite sulfur configuration. On the basis of this nomenclature system, the proportion of adducts obtained in each case can be straightforwardly related to the conformational preferences of the starting dienophile around the C-S bond. Thus, the relative stability of the rotamers becomes one of the main factors responsible (in many cases it will be the only one considered by the authors) for the stereochemical course of these cycloadditions.

2
Diels-Alder Reactions

Diels-Alder reactions have become one of the most widely used tools in asymmetric synthesis due to their high stereocontrol. Although this can be achieved by the use of chiral dienophiles, chiral dienes, or chiral Lewis acids [1], the vast majority of the work on asymmetric Diels-Alder reactions deals with the use of chiral dienophiles because they usually show higher and more predictable facial stereoselectivity. This is the reason that the number of papers concerning sulfinyl ethylenes is rather higher than those referring to sulfinyldienes. Otherwise the efficiency in the control of the π-facial selectivity of the Diels-Alder reactions is usually dependent on the distance between the chiral center and the dienophilic double bond (or dienic system). In this sense, alkenyl (and dienyl) sulfoxides have some advantages as compared with other dienophiles (or dienes) because the chiral sulfur atoms are directly joined to the reactive moieties in the cycloadditions. This section is divided into two parts concerning reactions with the sulfinyl group anchored to dienophile (Sect. 2.1) and to diene (Sect. 2.2) respectively. As criteria to subdivide each of these parts we have used the nature of the other functional groups present at sulfinyl dienophiles and the position of the sulfur function at the dienic system.

2.1 Sulfinyl Dienophiles

2.1.1 *α,β-Unsaturated Sulfoxides*

On the basis of the –I and –M effects assigned to the sulfinyl group, it is usually considered to be an activator of double bonds in their reactions with dienes. However, the low reactivity exhibited by racemic vinyl sulfoxides in Diels-Alder cycloadditions has decreased interest in these substrates, which have as a result been comparatively much less studied than other dienophiles such as acrylates or vinyl sulfones. Nevertheless, the chemical versatility of the sulfinyl group has allowed the use of vinyl sulfoxides as synthetic equivalents of many interesting dienophiles (poorly reactive or difficult to obtain) on the basis of simple reactions of the resulting adducts. Thus, 1-phenylsulfinyl ethylene has been used as a synthetic equivalent of ethylene [12] (based on the easy reductive cleavage of the C-S bond), acetylene [13], and allene [14] (pyrolytic elimination of the sulfinyl group), or ketene (*α*-silylation followed by sila-Pummerer rearrangement [15]) in Diels-Alder reactions (Scheme 1).

The first paper reporting the use of optically pure vinyl sulfoxides in an asymmetric Diels-Alder reaction (and therefore their use as a chiral equivalent of ethylene) was due to Maignan and Raphael [5]. In this paper, published in 1983, they studied the reaction of cyclopentadiene and (*R*)-(+)-*p*-tolyl vinyl sulfoxide (**1**). These compounds were heated at 110 °C in a sealed tube without solvent, affording, in a high yield, a mixture of the four possible adducts **2** (Scheme 2). Their stereochemistry was established by ^{1}H-NMR spectroscopy, and that of the *endo* and *exo* major products was confirmed by chemical correlation (reduction of the sulfinyl group followed by *α*-halogenation and oxidative hydrolysis) with known enantiomeric dehydronorcamphors. Several years later [16], a combined ^{1}H-NMR and X-ray study of three of the four possible racemic adducts obtained in reactions of cyclopentadiene with (±)-phenylsulfinyl ethylene allowed unequivocal assignment. The comparison of the NMR parameters of these adducts with those corresponding to compounds **2** (Scheme 2) reinforced the assignment made by Maignan and Raphael [5].

Despite the low reactivity and poor stereoselectivity of compound **1** as a dienophile, the main interest of the Maignan and Raphael's paper [5] derives from the fact that it was the first one involving the use of enantiomerically pure sulfoxides in Diels-Alder reaction, which would be used profusely later in asymmetric synthesis. For this reason it deserves some additional comments. From Scheme 2 can be deduced a moderate *endo* orientating character of the sulfinyl group [*endo*-adducts are the major ones (66%) in the mixture]. Although no explanation about the stereochemical behavior of compound **1** was offered in

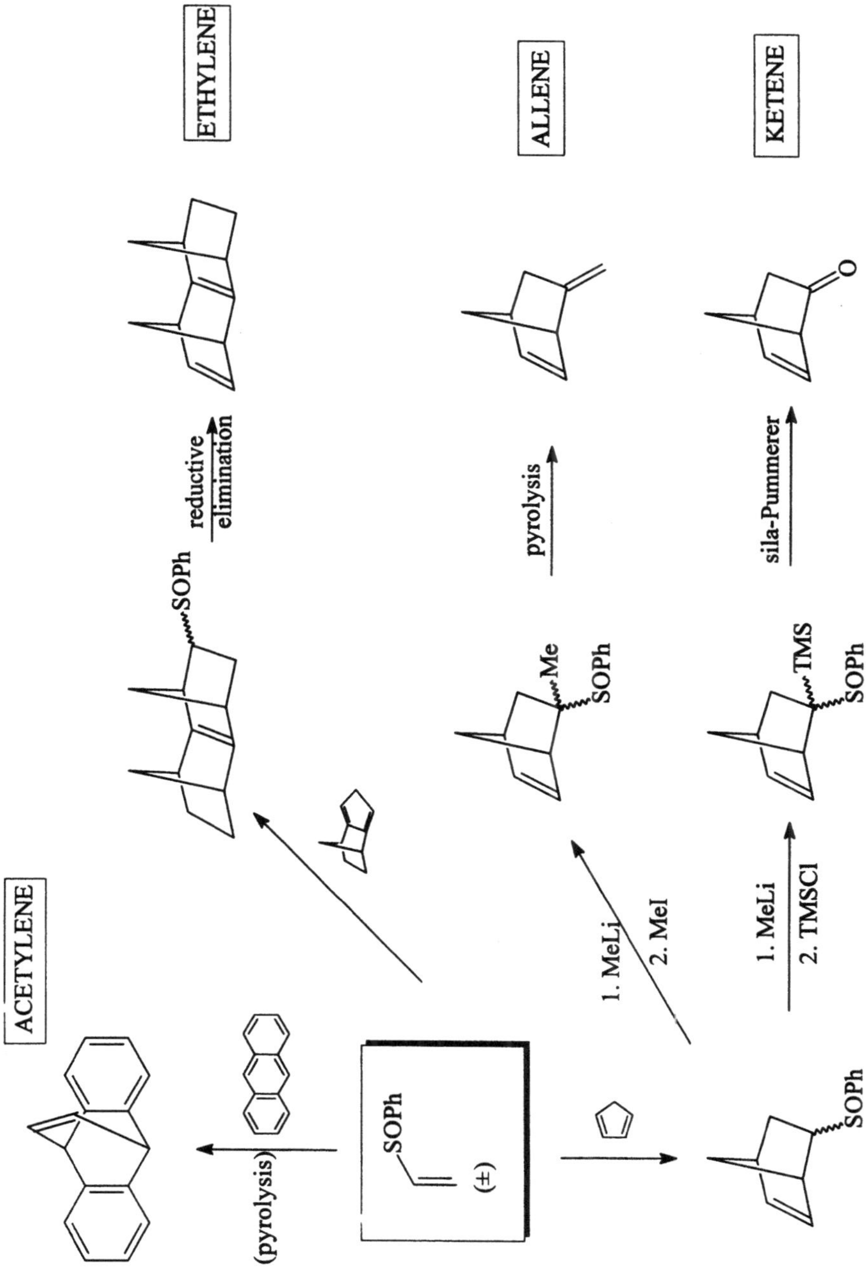
ACETYLENE
ETHYLENE
ALLENE
KETENE
SOPh
(±)
(pyrolysis)
reductive
elimination
pyrolysis
sila-Pummerer
1. MeLi
2. MeI
1. MeLi
2. TMSCl
Me
TMS
SOPh

Scheme 1

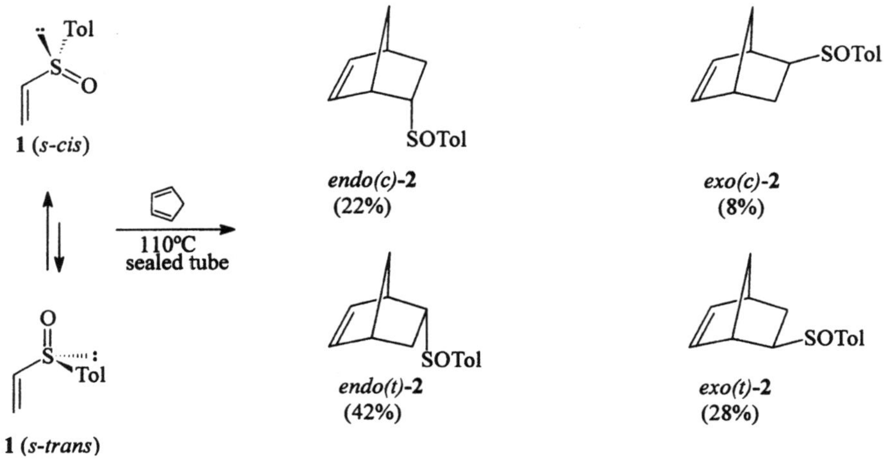

Scheme 2

the original paper, it has been further discussed by other authors. Thus, Arai and Koizumi suggested [10c] that the poor π-facial selectivity of the cycloaddition of **1**, in both *endo* (42:28) and *exo* (28:8) modes, points to a small difference of the two ground-state conformational energies of the dienophile, determining a significant contribution from both *s-cis* and *s-trans* rotamers (Scheme 2). This suggestion was based on the assumption that the π-facial selectivity of these cycloadditions ought to be controlled mainly by steric grounds and, thus, cyclopentadiene should attack from the less hindered lone-pair side in each mode. Cycloaddition of **1** in the *s-cis* conformation would afford the *endo*(*c*)-**2** (22%) and *exo*(*c*)-**2** (8%) adducts, while *endo*(*t*)-**2** (42%) and *exo*(*t*)-**2** (28%) would be the result of the diene attack on the *s-trans* conformation. The fact that *endo*(*c*) and *exo*(*c*)-adducts were obtained as the minor products is not clear because they derive from the *s-cis* conformation, which is the most stable one according to theoretical calculations [17]. It suggests that the conformational stability must not be the only factor considered in order to explain the stereoselectivity of these cycloadditions. This issue is discussed later.

Kahn and Hehre [18] postulated a different explanation. According to this, the π-facial selectivity of the Diels-Alder reactions of vinyl sulfoxides arises from the electrostatic preferences of a "nucleophilic" diene to avoid the electron rich lone pair on sulfur, even at the expense of encountering a sterically bulky substituent. In the case of the *p*-tolylsulfinyl ethylene, this proposal would involve favored attack from the more hindered face supporting the *p*-tolyl group (the opposite one to that shown in Fig. 1), thus avoiding electrostatic repulsion with the lone electron pair. According to this explanation the major adducts (*endo*(*t*)-**2** and *exo*(*t*)-**2**) would derive from the attack of diene on the dienophile adopting the most stable *s-cis* conformation. This proposal was criticized by Koizumi et al. on consideration of the conformational preferences of several optically pure substituted vinyl sulfoxides established from their X-ray and CD spectra [19]. Many other cases of evidence supporting the proposals of Koizumi et al. have been reported (see later) and thus the steric approach control is currently the most widely accepted by researchers working in the field.

The only paper concerning catalysis by Lewis acids of the Diels-Alder reactions of these simple sulfinyl ethylenes was due to Ronan and Kagan [20], who studied the influence of TMSOTf in the reaction of compound (S)-**1** with cyclopentadiene and furane. In the first case, the reaction occurs at 0 °C in 3 h, giving an 89:11 mixture of *endo* and *exo* adducts (overall yield 60 %) with very high π-facial selectivity (*de* > 92 %). The high efficiency of the catalyst increasing the reactivity of **1** also made possible its reaction with furan, which evolved with low *endo/exo* selectivity (55:45) and lower π-facial selectivity (*de* ≈ 70 %) than that observed with cyclopentadiene. These excellent results were nevertheless, eclipsed by those reported in the same paper [20] concerning the activation of

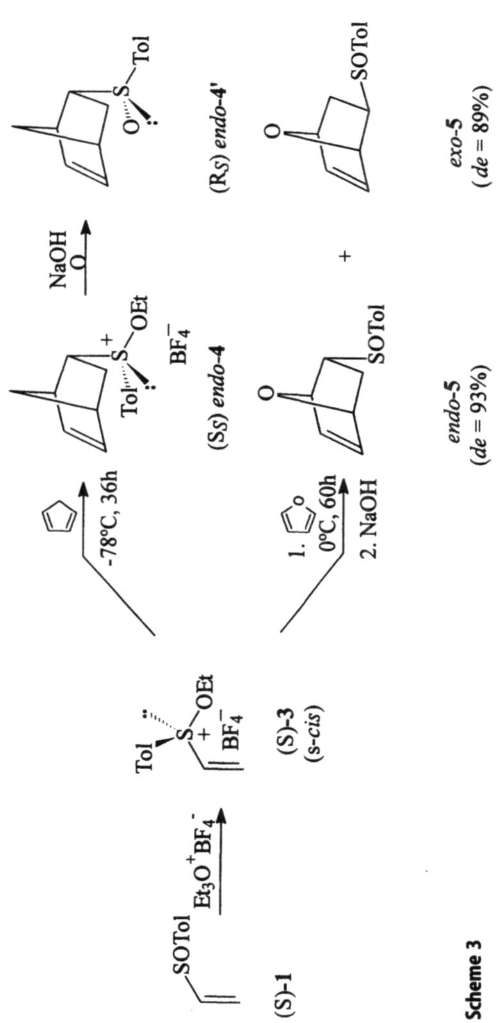

Scheme 3

compound (S)-**1** by Meerwein's reagent, which converted this sulfoxide into the alkoxysulfonium salt (S)-**3** (Scheme 3). This reacts with cyclopentadiene even at –78 °C (36 h) with a complete *endo/exo* and π-facial selectivity, yielding only one adduct, (S_S)-*endo*(*c*)-**4**, which was transformed into the sulfoxide (R_S)-*endo*-**4′** by treatment with NaOH with inversion of the configuration at sulfur (Scheme 3). A complete control of stereoselectivity was also observed in reactions with 2,3-dimethylbutadiene (40 % yield), but other dienes, including cyclohexadiene and anthracene, did not react.

Compound (S)-**3** reacts with furan at 0 °C yielding a mixture of *endo*-**5** (*de* = 93 %) and *exo*-**5** (*de* = 89 %) adducts in 90 % yield (Scheme 3), evidencing a very high π-facial selectivity in both *endo* and *exo* approaches. However, the reaction occurs in a low *endo*-selective manner (*endo/exo* = 59 : 41). The absolute configuration of the resulting adducts was not unequivocally established. By decreasing the reaction temperature, the π-facial selectivity becomes higher, but the observed yields were quite poor. In contrast with the high reactivity exhibited by compound **3**, other alkenyl sulfoxides bearing alkyl groups at the double bond were not able to react even with cyclopentadiene (in the best case, 25 % yield of a mixture of adducts was obtained).

Despite the excellent results obtained from **3**, only one additional paper has been reported concerning its use in asymmetric synthesis. It describes the reaction of (*R*)-**3** with 5-benzyloxymethyl-1,3-cyclopentadiene to afford almost exclusively one adduct, which was transformed into the sulfoxide (S_S)-**6** by reaction with NaOH [21]. Bicyclic compound **6** was converted into enantiomerically pure norbornenone **7**, a key intermediate in Corey's syntheses of prostaglandins (Scheme 4).

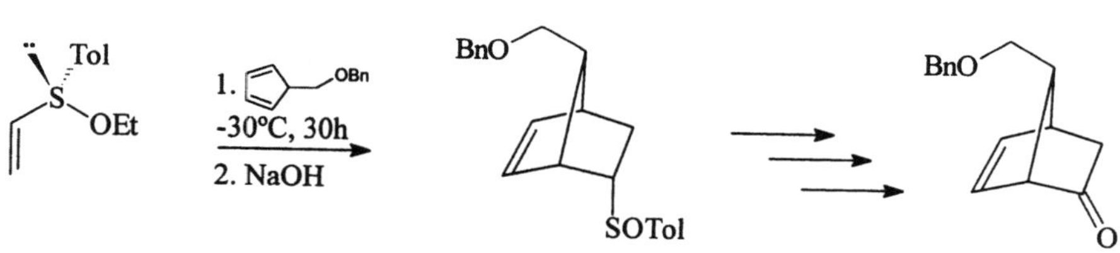

(R)-**3** (S_S)-**6** **7**

Scheme 4

Although the strong increase in the reactivity, promoted by formation of the alkoxysulfonium salt, could be expected, the intriguing change in the stereoselectivity (which is low for sulfoxide **1** but almost complete for alkoxysulfonium salt **3**) remains unexplained. The stereochemistry of the major adducts obtained in these reactions was rationalized by Ronan and Kagan [20] by assuming a steric approach control of the diene toward the alkoxysulfonium salt, which adopts the conformation with the S-OEt bond in an *s-cis* arrangement, thus resulting in *endo*(*c*)- adducts. Nevertheless, no explanation is offered relating to the factors determining the conformational preferences restrictions of compound **3** or the higher reactivity of the favoured conformation. The lack of reactivity of other alkoxysulfonium salts different from **3** was also not explained.

Optically pure alkynyl sulfoxides have also been studied as dienophiles in asymmetric Diels-Alder reactions. In 1988, Maignan and Belkasmioui [22] and Lee et al. [23] independently reported the behavior of several sulfinyl acetylenes. Although the paper from Lee et al. presents a wider Diels-Alder study, it was focused on racemic sulfoxides. It described the synthesis of (±)-*p*-nitrophenylsulfinyl ethyne and (±)-1-*p*-nitrophenylsulfinyl propyne, and their reactions with six different dienes (cyclic and acyclic), its most interesting finding being the high reactivity of these dienophiles in contrast to the rather low reactivity of vinyl sulfoxides. Thus, reactions of alkynyl sulfoxides with cyclopentadiene and cyclohexadiene took place in 6 h at room temperature or in refluxing benzene, respectively, while with other less reactive dienes the reactions were carried out in refluxing xylene (anthracene or tetraphenyl cyclopentadienone) or in sealed tubes at 140 °C (acyclic dienes). Regioselectivity in reactions with isoprene was non-existent (a 1:1 mixture of regioisomers was obtained), and the stereoselectivity of the reactions was not studied (mixtures of diastereoisomers were reported in some cases, but their relative configurations were not determined).

The paper from Maignan and Belkasmioui [22] described the synthesis of the optically pure ethynyl *p*-tolylsulfoxide **8a** and its trimethylsilyl derivative **8b**, as well as their reactions with cyclopentadiene, including stereochemical studies to assign the configuration of the two diastereoisomers obtained (Scheme 5). The observed stereoselectivity was not explained.

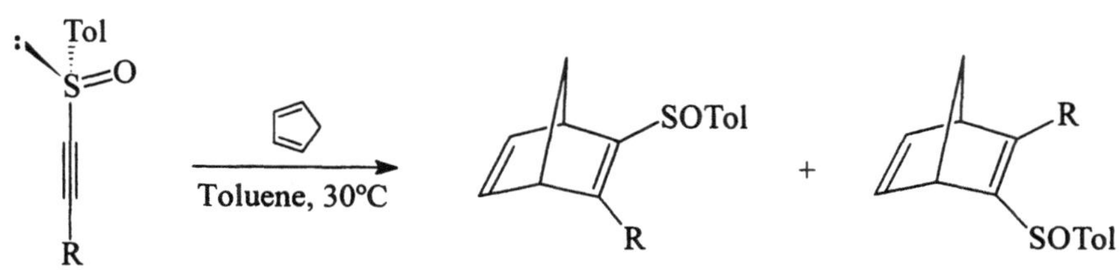

8a (R = H)
8b (R= TMS)

(70) (30)

Scheme 5

Reactions of compound **8a** with butadiene, 2,3-dimethylbutadiene, and isoprene required more drastic conditions (150 °C, autoclave) [24]. In the case of isoprene, a 60:40 regioisomeric mixture was obtained (the major product being the *para*-adduct). The influence of Lewis acid catalysis on the reaction of cyclopentadiene with different ethynyl aryl sulfoxides [Ar = *p*-Tolyl, *o*-nitrophenyl, and 1(2-methoxynaphthyl)] has been recently reported [25]. Dienophilicities of acetylenic sulfoxides were greatly enhanced by the presence of various Lewis acids (the effect being greater with stronger Lewis acid), but the stereoselectivity was scarcely affected by the catalysis.

Two remarkable facts, neither of them considered by the authors in the original papers, must be taken into account. The first concerns the significant stereoselectivity observed in these reactions (*de* = 40%, Scheme 5), which suggests strong conformational restrictions around the C-S bond. The second one is related to the reactivity, which is similar for alkynyl sulfoxides **8a** and **8b** (despite the presence of the activating Me_3Si group in the latter), and much higher than that exhibited by sulfinyl ethylenes. As steric effects cannot be invoked to explain the conformational restrictions in alkynyl sulfoxides, it is necessary to assume that electronic interactions between the sulfinyl group and the π-system must be responsible for such restrictions. Moreover, these interactions could be dependent on the substituent at the triple bond and therefore could be related to the relative reactivity of **8a** and **8b**. We will return later to these points.

From the above results it can be inferred that sulfinyl ethylenes are not good dienophiles because of their low reactivity and poor stereoselectivity. In order to improve both of these factors, different electron-withdrawing groups (able to increase dienophilic reactivity and to restrict conformational movement around the C-S bond, thus improving the stereoselectivity) were incorporated into the double bond. They are presented in the sections below, classified according to the number and kind of the additional activating groups present in the vinyl sulfoxide moiety.

2.1.2
Monoactivated Vinyl Sulfoxides

2.1.2.1
Alkoxycarbonyl Derivatives

Many electron-withdrawing groups have been incorporated into the dienophilic double bond of vinyl sulfoxides, alkoxycarbonyl being the most widely studied. Concerning the behavior of optically pure 3-sulfinylacrylates, the results of three different research groups were published almost simultaneously in 1984. Chronologically, the first one was that from Koizumi's group [26], which reported the synthesis of (*Z*)- and (*E*)-ethyl (*R*)-2-methyl, 3-*p*-tolylsulfinyl acrylates (**9a** and **10a**, Scheme 6) and their reactions with cyclopentadiene. The yields obtained in the synthesis of the dienophiles were poor (lower than 20%) and their reactivity low (5 h were required in a sealed tube at 90 °C). The most interesting finding of these cycloadditions was the high π-facial selectivity, which became almost complete both in the *endo* and *exo* approaches starting from the (*Z*)-isomer (**9a**). In the case of compound **10a**, the π-facial selectivity was lower in the *endo* approach ($de \approx 62\%$, Scheme 6). The *endo/exo* selectivity was moderate or low with both substrates.

Almost simultaneously, Maignan's group reported the synthesis and reactivity of optically pure (*Z*)-methyl (*R*)-3-*p*-tolylsulfinyl acrylate (**9b**) with cyclopentadiene [27]. This dienophile shows higher reactivity (60 h in toluene at 4 °C) and better *endo/exo* selectivity (92:8) than **9a**, exhibiting an almost complete π-facial selectivity for the *endo* and *exo* approaches (Scheme 6). In 1986 Maignan et al. reported the results obtained in the Diels-Alder reaction of cyclopentadiene with the (*E*)-isomer **10b** [28]. This dienophile provided very low *endo/exo* and π-facial stereoselectivities, yielding a mixture of the four possible adducts, all of them in significant amounts. The main conclusion drawn from this paper is the fact that the *endo*-orientating character of the sulfinyl and ethoxycarbonyl

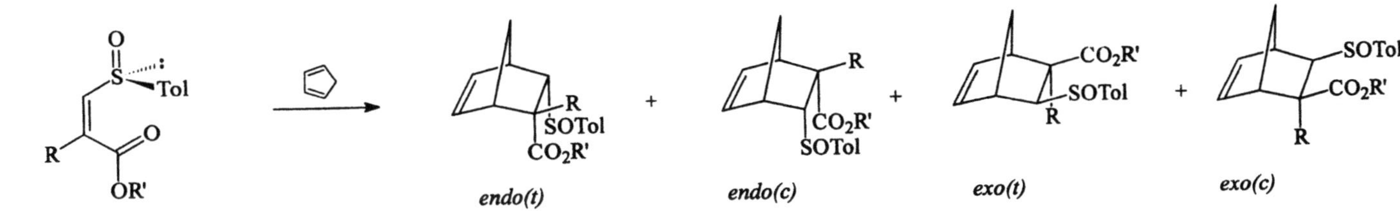

Compound	Conditions				
9a (R=Me, R'=Et)	90°C, 5h, sealed tube	63	2	35	-
9b (R=H, R'=Me)	4°C, 60h	92	-	8	-
9c (R=H, R'=Bu)	Benzene, reflux, 2h	77	3	11	9

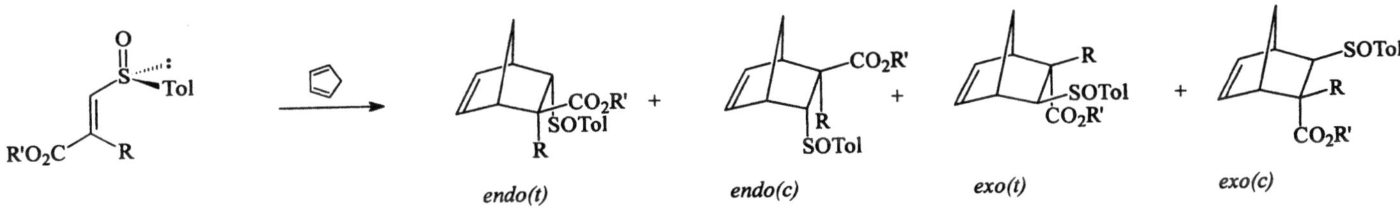

Compound	Conditions				
10a (R=Me, R'=Et)	90°C, 5h, sealed tube	63	15	22	-
10b (R=H, R'=Me)	4°C, 60h	42	12	30	16
10c (R=H, R'=Bu)	Benzene, reflux, 2h	37	16	42	15
	BF_3OEt_3	56	22	19	3
	$SnCl_4$	23	3	55	19

Scheme 6

groups seems to be quite similar. The fact that the π-facial selectivity was lower in reactions of **10b** than that observed for the other substrates suggested that it must be due to the conformational restrictions around the C-S bond existing in **9a**, **10a**, and **9b** but not in **10b** [29].

In 1985 Davis and Brimble [30] reported the results obtained in the reactions of cyclopentadiene with butyl esters **9c** and **10c**, which are quite similar to those reported by Maignan et al. (Scheme 6). The (*Z*)-isomer mainly gave a 77:11 mixture of *endo* (*t*) and *exo* (*t*) adducts, whereas the (*E*)-isomer yielded a mixture of the four possible adducts. The addition of Lewis acids to the reaction of the (*E*)-isomer improved the reactivity of dienophile **10c**, but its diastereoselectivity, which was significantly changed, remained rather low and insufficiently attractive for synthetic purposes.

In order to explain the stereochemical course of these reactions, Koizumi assumes that steric grounds govern the approach of cyclopentadiene to dienophiles **9a** and **10a**. Once more, the favored attack of the diene will take place from the diastereotopic face of vinyl sulfoxide supporting the smallest substituent at sulfur [26]. According to this assumption, the π-facial diastereoselectivity must be dependent on the conformational preferences of the dienophile around the C-S bond. Therefore, as the stereochemistry of the major adducts is *endo*(*t*) and *exo*(*t*) (Scheme 6), Koizumi proposes that the precursor dienophiles **9a** and **10a** must adopt mainly the conformation with the sulfinyl oxygen in an *s-trans* arrangement with respect to the double bond (A rotamer in Fig. 2), and therefore the favored approach of cyclopentadiene would take place from the bottom face of dienophile, which supports the lone pair electrons (much smaller than the tolyl group). In the case of **9a**, this conformational preference can easily be justified as minimizing the electrostatic repulsion between the sulfinyl and alkoxycarbonyl oxygens, but it is not as obvious for **10a**, where steric interactions must be considered exclusively. The predominance of a conformation such as **B-10a** (Fig. 2), exhibiting minimum steric interactions, would also explain the stereochemical course of the reaction (the favored attack would take place from the bottom face supporting the smallest substituent). The differing magnitude of the π-facial selectivity observed for the evolution of **9a** (*de*>94%) and **10a** (*de*~61%) could also be explained taking into account the spatial arrangement of the substituents in conformations **A-9a** and **B-10a**, with the latter exhibiting a lower steric differentiation between its diastereotopic faces.

The two major adducts obtained in the reaction of **9a** with cyclopentadiene were used by Koizumi et al. in the synthesis of bicyclic sesquiterpenes [31]. From

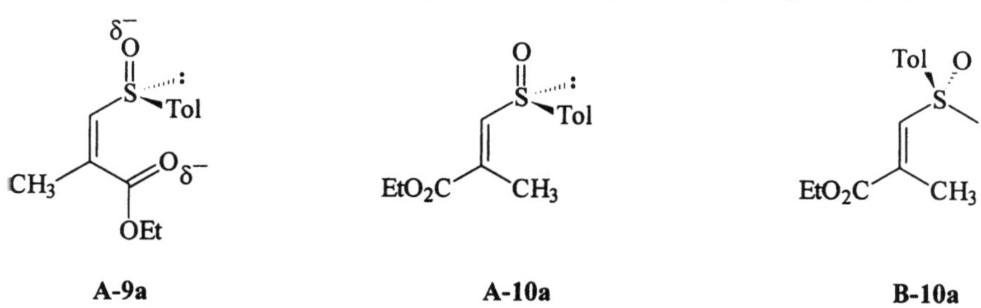

Fig. 2

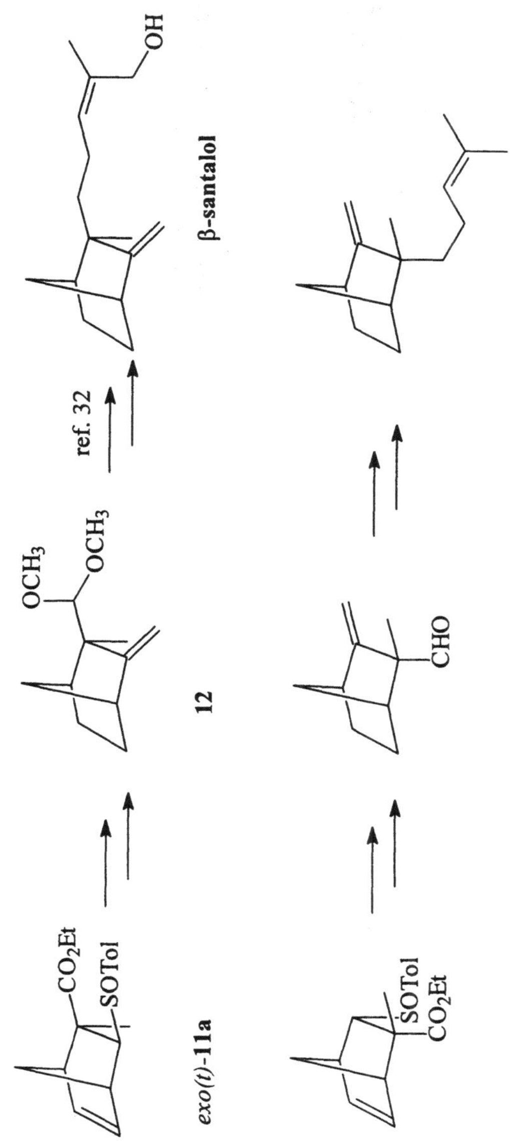

exo(t)-**11a** → **12** → β-santalol

endo(t)-**11a** → **13** → epi- β-santalene

Scheme 7

exo(*t*)-**11a** he prepared the optically pure acetal **12**, the racemic form of which had been previously used in the synthesis of (±)-*β*-santalol [32]. Furthermore, the *endo*(*t*)-**11a** adduct was transformed into aldehyde **13**, which was further converted into optically pure (+)-*epi*-*β*-santalene in 11 steps (Scheme 7).

Two main problems were associated with the use of optically pure 3-*p*-tolylsulfinyl acrylates as dienophiles. The first was due to the high number of reaction steps and consequently low overall yields involved in their preparation, and the second to the fact that these dienophiles were insufficiently reactive with a variety of dienes. In order to solve the first problem, de Lucchi et al. [33] designed new sulfoxides bearing bornyl and isobornyl moieties joined to the sulfur atom (instead of the *p*-tolyl group). This allowed them to develop a new strategy to obtain the sulfoxides by asymmetric oxidation of thioethers, taking advantage of their chiral moieties. The reactions of these new dienophiles with cy-

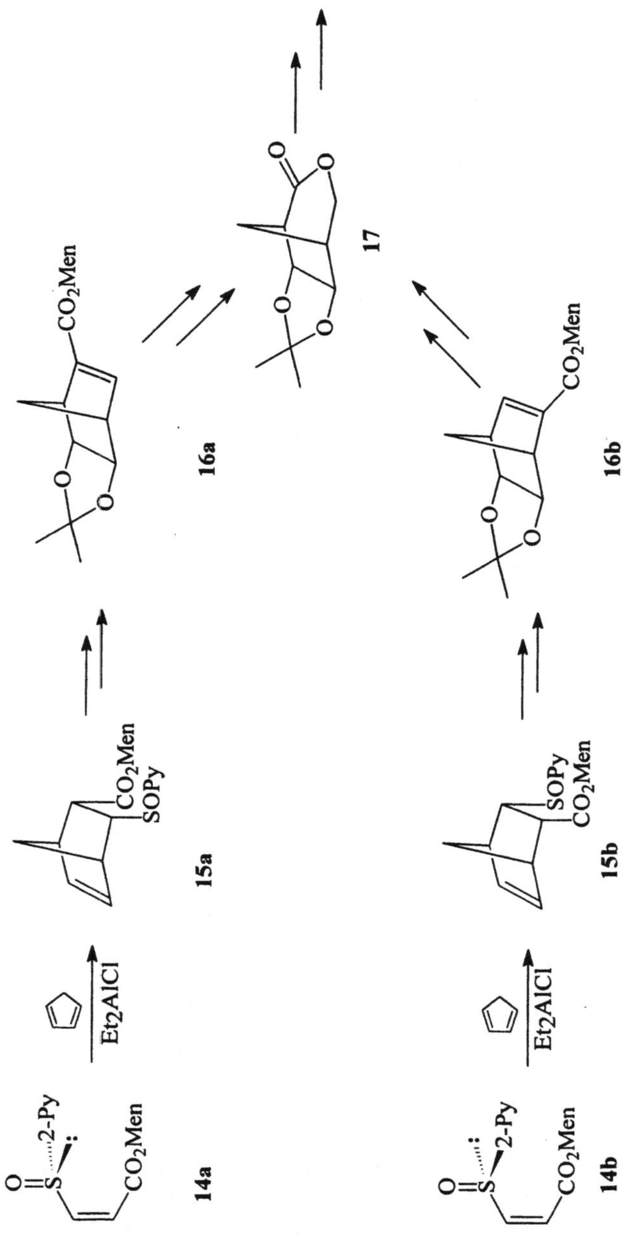

Scheme 8

clopentadiene yielded only the *endo*-adduct starting from the (*Z*)-sulfinyl acrylate, but a complex mixture of adducts from its (*E*)-isomer.

The rather low reactivity of the 3-*p*-tolylsulfinyl acrylates (they do not react with furan even under forcing conditions) prompted the search for more reactive dienophiles. In this context, pyridylsulfinyl derivatives proved to be more efficient than the arylsulfinyl ones. Thus, menthyl-3-(2-pyridylsulfinyl)propenoates **14a** and **14b** were prepared from (+)-menthyl propiolate in low yields [34]. Their reactions with cyclopentadiene proceed smoothly in the presence of Et_2AlCl at −70 °C to afford just one *endo* diastereoisomer **15a** or **15b** [35] (in the absence of the catalyst the π-facial selectivity for the *endo* approach was lower than that observed for the *p*-tolylsulfinyl derivatives [10c]). These compounds were transformed into **16a** or **16b** [36] respectively (Scheme 8), both allowing the synthesis of the bicyclic lactone **17** (known as Ohno's lactone), a key intermediate in Ohno's synthesis of (−)-aristeromycin and (−)-neplanocin A [37].

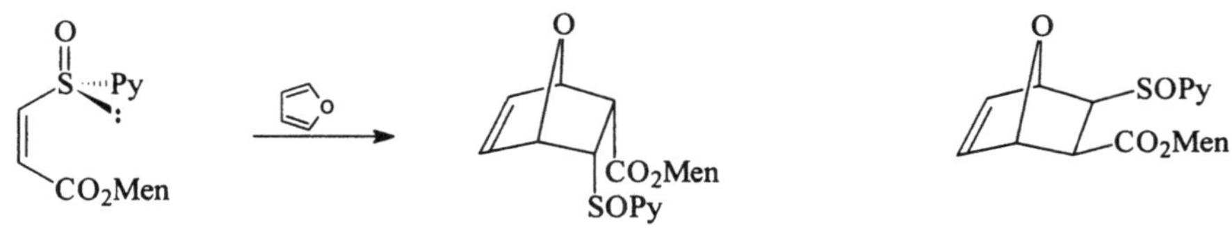

14a		*endo(t)*-**18a**	*exo-(t)*-**18a**
	Et_2AlCl, rt, 7 days	44% (93:7)	25% (96:4)
	50°C, 6 days	49% (55:45)	13% (92:8)

Scheme 9

Pyridylsulfoxides **14a** and **14b** are able to react with furan, yielding a mixture of four adducts **18**. In Scheme 9, only the major *endo*(*t*)- and *exo*(*t*)- adducts, derived from **14a**, have been depicted. In the presence of Et_2AlCl the reaction of **14a** required 7 days at room temperature to reach completion [34], and its π-facial selectivity was very high (93:7 mixture of *endo* adducts and 96:4 mixture of *exo*-adducts), the results obtained from **14b** being similar. In contrast, the *endo/exo* selectivity is only moderate for both dienophiles (<2:1). In the absence of the Lewis acid (50 °C, 6 days), the *endo/exo* selectivity of these reactions was slightly higher (almost 4:1 for **14a**), but the π-facial selectivity for the *endo*-approach was clearly poorer (55:45).

The adducts resulting from Diels-Alder reaction of pyridyl sulfoxides with furan have been used in the synthesis of a number of natural products. Thus, a new procedure for the total synthesis of optically actives C-nucleosides was reported by Koizumi et al. [38], who prepared D-showdomycin (**19**) and (D)-3,4-*O*-isopropylidene-2,5-anhidroallose (**20**) from the *endo*(*t*)-**18a** adduct (Scheme 10). (+)-Methyl 5-epishikimate (**21**) [39] and pentaacetyl-β-D-mannopyranose (**22**) [40] were also obtained starting from *endo*(*t*)-**18a** (Scheme 10), the cleavage of the oxygenated bridge being the key step of these transformations.

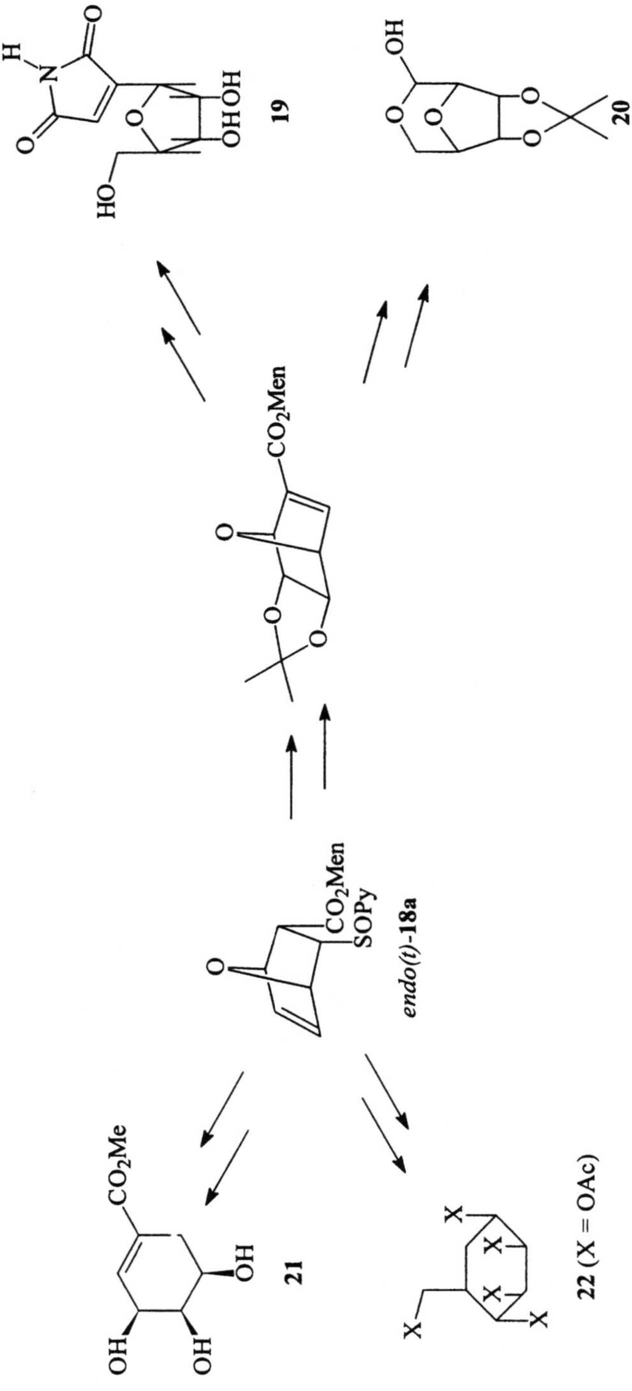

Scheme 10

The results obtained from substituted furans are very similar. Thus, reaction of **14a** with 3,4-dibenzyloxyfuran [41] yielded a 63:37 mixture of *endo* (*de*>92%) and *exo* (*de*>94%) adducts [40]. After separation and purification, the major adduct *endo*(*t*)-**23** was used to prepare pentaacetylpseudo-α-*l*-mannopyranose **24** (Scheme 11).

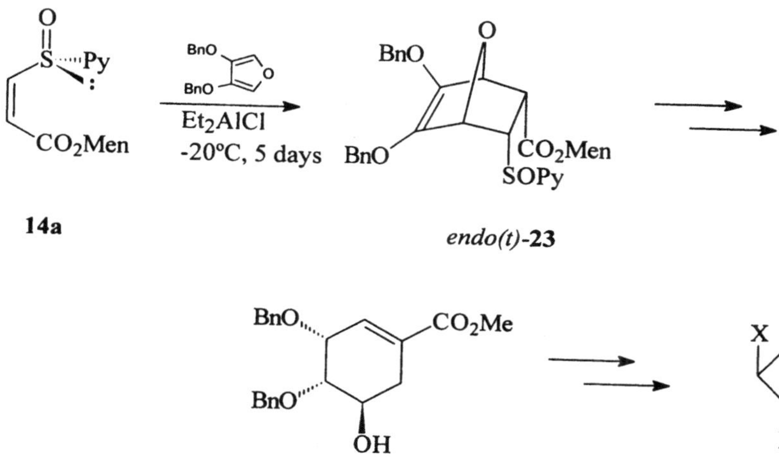

14a *endo(t)*-**23**

Scheme 11 **24** (X = OAc)

The introduction of an electron-withdrawing group (NO_2 or CF_3) at the pyridine ring of 2-pyridyl sulfoxides enhances the dienophilic reactivity. The results obtained from racemic dienophiles (±)-**14**, (±)-**25**, and (±)-**26** with furan are shown in Scheme 12 [41]. The reactions took place with moderate *endo*/*exo* selectivity in all cases but the π-facial selectivity was quite high (for both *endo* and *exo* approaches) when substituted pyridines **25** and **26** were used as dienophiles.

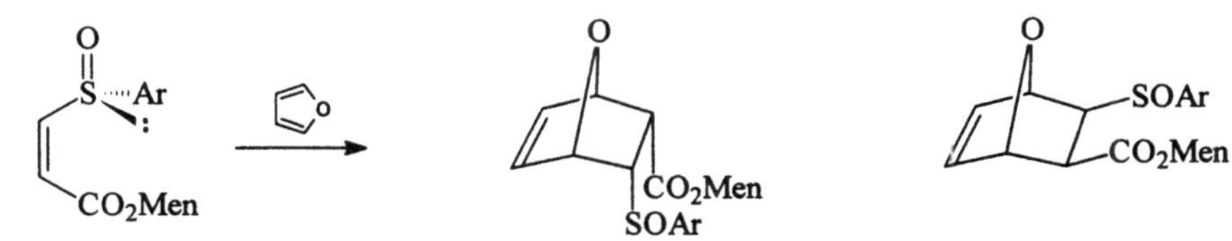

14a *endo(t)* *exo(t)*

Ar = 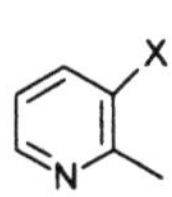

	endo(t)	*exo(t)*
(±)-**14** (X=H)	49 (55:45)	13 (93:7)
(±)-**25** ($X=NO_2$)	56 (>98:2)	16 (>98:2)
(±)-**26** ($X=CF_3$)	63 (>92:8)	7 (93:7)

Scheme 12

The stereochemical results of all these reactions were explained by assuming as predominant the conformation having the sulfinyl oxygen in an *s-trans* arrangement [42] but the reasons justifying the significant differences in π-facial selectivity for substrates **14**, **25**, and **26** were not discussed.

The obtaining of the optically pure (S_S)-**26** was achieved in 20% yield by crystallization from the 1:1 mixture of epimers at sulfur resulting from the oxidation of the corresponding thioether. The reaction of (S_S)-**26** with 2-methoxyfuran (6 days, 0 °C, toluene) afforded only compound **27** (in a complete regio, *endo*, and π-facial selective manner), which was transformed into glyoxalase I inhibitor **28** [42] in a seven-step sequence (Scheme 13).

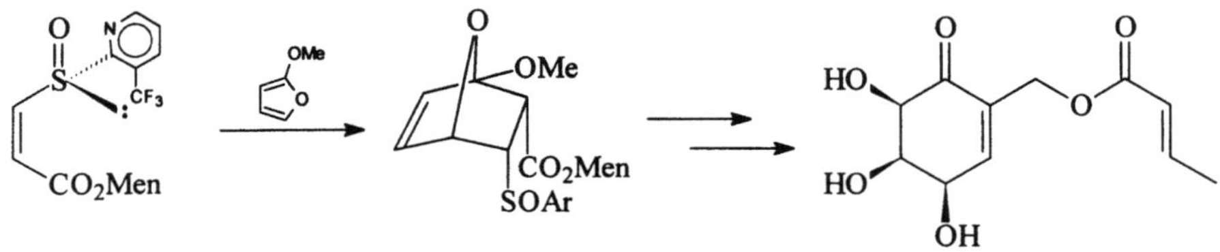

(S_S)-**26** **27** **28**

Scheme 13

Recently, Koizumi et al. reported that the use of high pressures allows exclusion of Lewis acids (which are not compatible with many dienes and/or adducts) in the cycloadditions of acrylates [43]. Thus, methyl 3-alkylsulfinyl acrylate **29** is able to react with cyclopentadiene, furan, and 2-methoxyfuran at 12.6 Kbar (Scheme 14). Both π-facial and *endo*/*exo* selectivities are very high in reactions with cyclopentadiene (only one adduct was obtained), whereas with furan derivatives the *endo/exo* selectivity is clearly lower. In reactions with cyclopentadiene it could be established that high pressures do not have a significant influence on the diastereoselectivity. The transformation into (–)-COTC of the major adduct obtained from 2-methoxyfuran was carried out in order to confirm its absolute configuration.

Ketene acetal sulfoxide **30** has been used as a chiral ketoester ketene acetal equivalent, because it undergoes a ready enantiocontrolled reaction with cyclopentadiene at –78 °C in the presence of BF_3 yielding a 96:4 mixture of *endo* and *exo* adducts with complete π-facial selectivity (Scheme 15) [44]. The *endo* selectivity decreased with other catalysts, but the π-facial selectivity remained complete, whereas under thermal conditions (139 °C, 15 h) a mixture of the four possible adducts was obtained. The adduct **31** was transformed into (+)-(1*R*,4*R*)-norbornenone in a four-step sequence.

The first report on the use of 2-sulfinyl acrylates as optically pure dienophiles was due to Koizumi et al. [45], who synthesized (+)-ethyl 2-*p*-tolylsulfinyl acrylate (**32a**) and studied its behavior as a dienophile. It reacted with cyclopentadiene in 6 h at room temperature yielding a mixture of the four possible adducts (Scheme 16). Significant π-facial selectivity (favoring *endo*(*c*) and *exo*(*c*) adducts) but moderate *endo/exo* selectivity were obtained. Milder conditions were required in the presence of $ZnCl_2$ (3 h at 0 °C), which inverted and substantially increased the π-facial selectivity (it became almost complete, favoring *endo*(*t*) and *exo*(*t*) adducts), but had only a small effect on the *endo* selectivity. (Scheme 16). The use of ZnI_2 instead of $ZnCl_2$ increases the reactivity (1 h at –20 °C) and improves the *endo/exo* ratio (*de* = 74%), as was recently reported for dienophile **32b** [46].

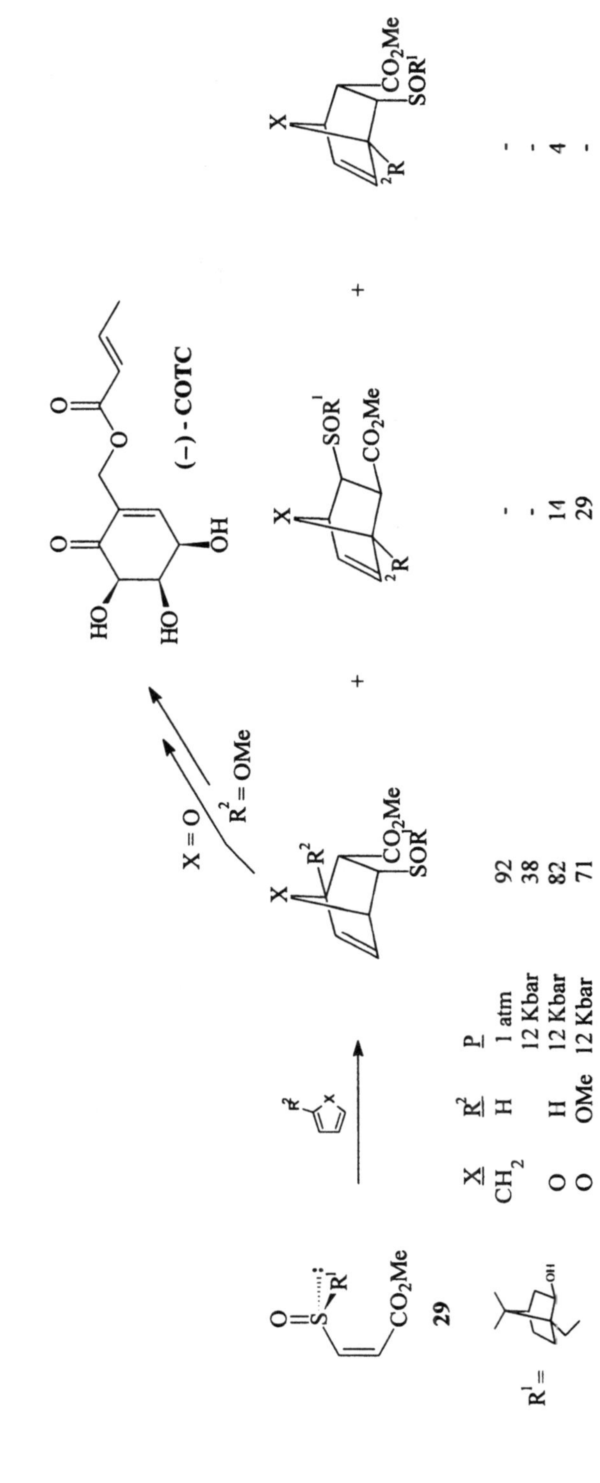
29
R^1 =
X = O
R^2 = OMe
(−) - COTC
X R2 P
CH2 H 1 atm 92 - -
12 Kbar 38 - -
O H 12 Kbar 82 14 4
O OMe 12 Kbar 71 29 -

Scheme 14

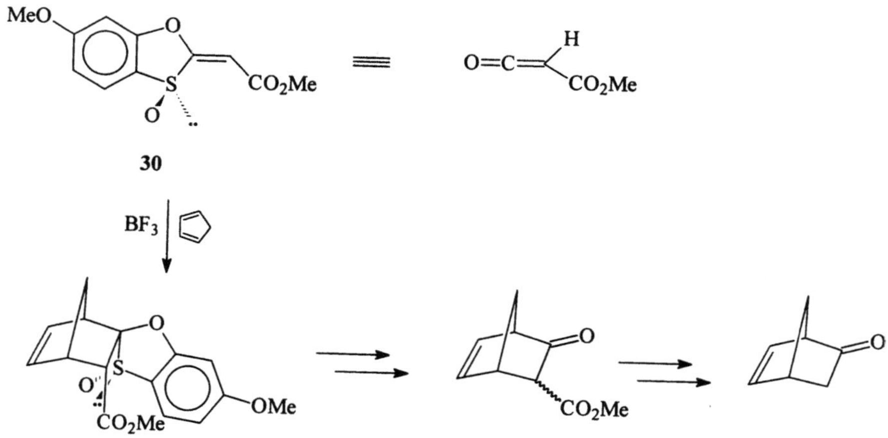

Scheme 15

On the assumption that cycloaddition is governed by a steric approach control, this behavior was explained by assuming that the favored rotamer of the dienophile changes with the reaction conditions. In the absence of catalyst, the sulfinyl oxygen adopts the *s-cis* arrangement with respect to the double bond (A in Fig. 3), whereas the addition of $ZnCl_2$ shifts the conformational equilibrium around the C-S bond toward the *s-trans* rotamer, which will form the chelated species B (Fig. 3). In the two cases the less-hindered face of the dienophile (that displaying the lone electron pair at sulfur) is opposite, which would justify the observed inversion of the π-facial selectivity (reactions on species B afford *endo*(*t*)-**33** and *exo*(*t*)-**33** as the major products, whereas *endo*(*c*)-**33** and *exo*(*c*)-**33** are predominant when diene attacks species A). The fact that conformational restrictions would be more severe in the presence of the chelating agent is consistent with the higher π-facial selectivity observed for these catalyzed reactions. According to these results, reactions of **32a** with anthracene catalyzed by $ZnCl_2$ (room temperature, 51 h) gave only one adduct (*endo*(*t*)- and *exo*(*t*)-adducts are now identical) [45]. Analogous results were recently obtained by Chau et al. [47] from (–)-menthyl 2-(*ortho*-methoxyphenylsulfinyl) acrylate. The higher reactivity of this dienophile (it requires 18 h at 0 °C in the presence of 3 equiv. of $ZnCl_2$) with respect to that of **32a** is attributed to the formation of a tridentate chelation involving methoxy, sulfinyl, and alkoxycarbonyl oxygens.

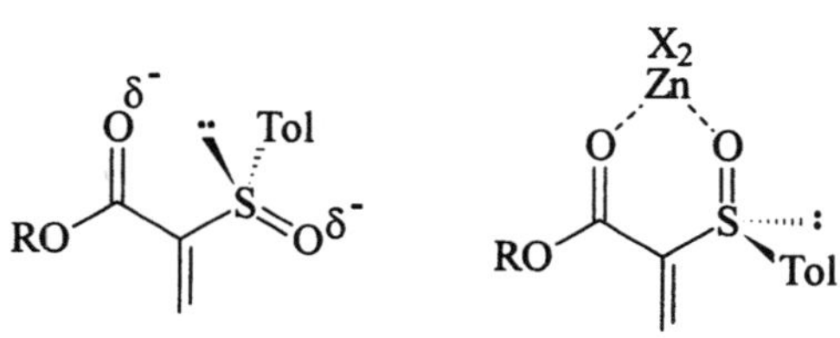

Fig. 3 **A** *(s-cis)* **B** *(s-trans)*

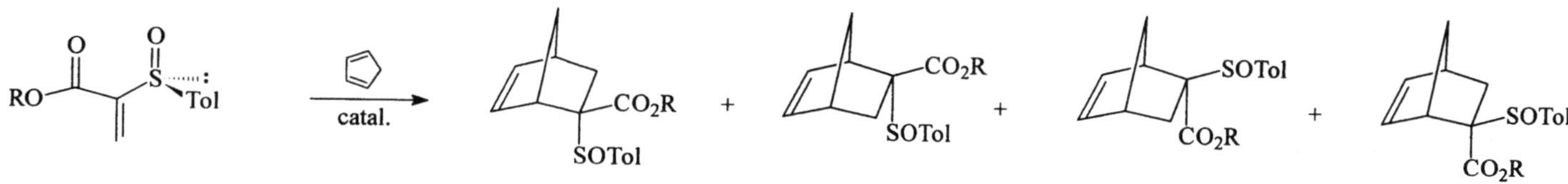

32a (R = Et)
32b (R= Bn)

endo(c)-**33** | *endo(t)*-**33** | *exo(c)*-**33** | *exo(t)*-**33**

Substrate	Catal.	Conditions	*endo(c)*-**33**	*endo(t)*-**33**	*exo(c)*-**33**	*exo(t)*-**33**
32a	-	rt, 6h	23	2	64	11
32a	$ZnCl_2$	0°C, 3h	2	19	2	77
32b	ZnI_2	-20°C, 1h	-	13	-	87

Scheme 16

Reactions of **32a** with acyclic dienes (butadiene and piperylene) yielded the cyclohexadienes derived from desulfinylation of the resulting adducts, precluding the deduction of any information about the stereochemical course of these reactions [45]. From the results of the reaction of racemic sulfoxide **34** with the acyclic diene **35**, Alexandre et al. [48] were able to isolate and characterize bicyclic adduct **36** as the only adduct because its desulfinylation required 30 min in CH_2Cl_2 under reflux (Scheme 17). This result proved that the *exo*-selectivity of these reactions with acyclic dienes is much higher than that with cyclic dienes. Moreover, this paper shows that the reactivity of 3-sulfinyl acrylates is lower than that of the 2-sulfinyl derivatives, because the former are not able to react with **35**.

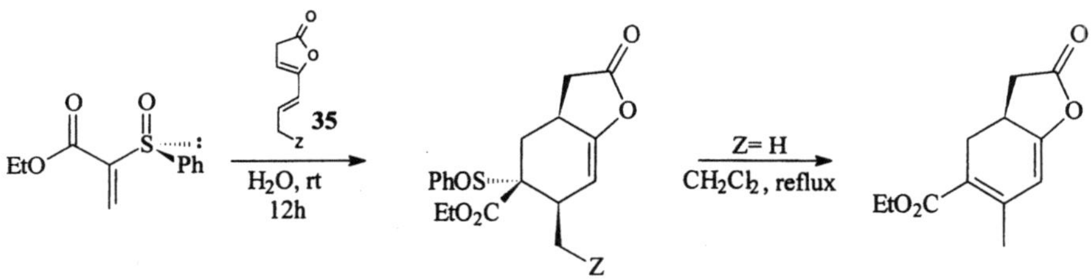

(±)**34**

36
Z= H, OTHP

Scheme 17

These results on racemic sulfinyl acrylates clearly suggested that the moderate *endo/exo* selectivity observed in reactions with cyclopentadiene (the main problem associated with the use of these dienophiles) could be substantially improved in reactions with acyclic dienes. Nevertheless, studies of the reactions of these dienes with optically pure dienophiles were not published until 1997 [46]. The paper concerning this topic reports the reaction under mild conditions (17 h at –25 °C) of Dane's diene with (*R*)-benzyl 2-*p*-tolylsulfinyl acrylate **32b**, catalyzed by $TiCl_4$ (Scheme 18). The NMR spectra of the crude reaction mixture revealed the formation of just one adduct (presumably the *exo*(*t*)-**37** as a consequence of the chelation of the dienophile) which suffered spontaneous desulfinylation at room temperature (complete after 36 h), affording cyclohexadiene **38** in 88% yield. In the absence of a catalyst, the reaction required the use of high pressures (12 Kbar, 24 h), yielding desulfinylated compound **40** in 72% yield. Adduct **39** was not detected. The optical purity of **40** (*ee* > 97%) indicated that the π-facial selectivity of the cycloaddition was completely controlled by the sulfinyl group. Although the absolute configuration of **40** was not unequivocally established, the stereochemical course of the reactions of **32b** with cyclopentadiene suggested it must be that depicted in Scheme 18. Compound **40** would result by desulfinylation of the adduct *exo*(*c*)-**39**, formed by approach of Dane's diene toward the less hindered face of the *s-cis* conformation of the dienophile.

The observed regioselectivities were opposit under the two conditions studied (see Scheme 18), which was explained by assuming that, in the absence of a

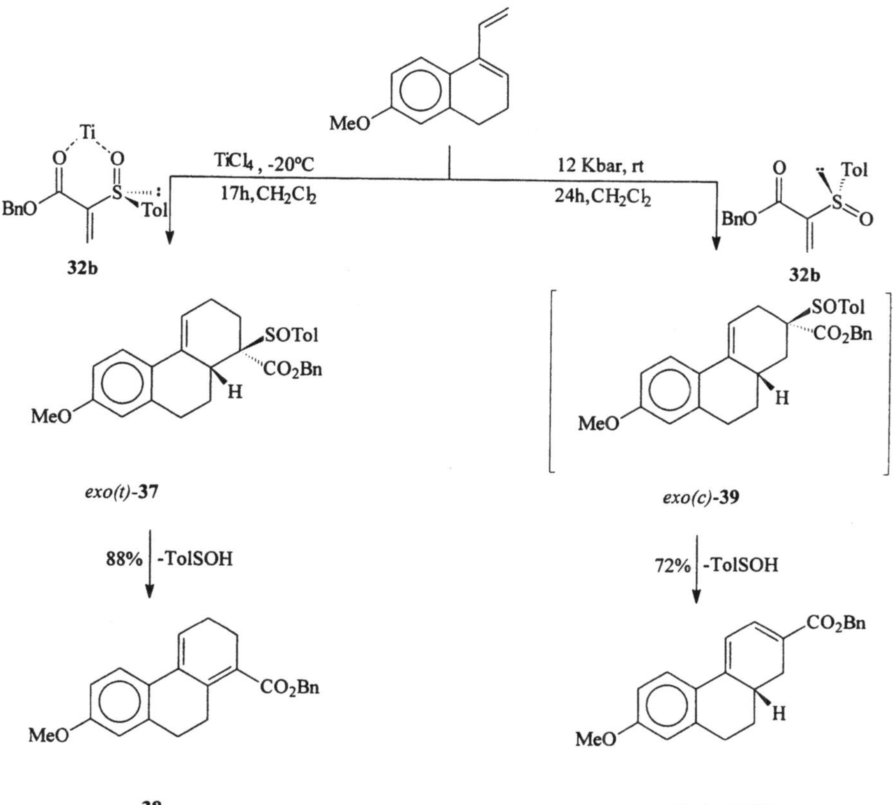

Scheme 18

Lewis acid, the orientating power of the aromatic fragment at C-2 is clearly higher than that of the methylene group at C-1. The inversion of the regioselectivity observed when the reaction was carried out in the presence of $TiCl_4$ was attributed to the association of the catalyst with the OMe group decreasing the electronic density of the aromatic ring and minimizing its orientating power. Nevertheless, the fact that these changes in the regioselectivity of cycloadditions to Dane's diene had been observed with this and other acyclic esters (see below), but not with sulfinyl lactones and sulfinyl cycloalkenones, suggests a more complex explanation which requires additional research.

Contrasting with the large number of papers concerning reactions of 3-sulfinyl acrylates with furan derivatives, as a key step in the synthesis of many interesting natural products (see above), only two papers have been published concerning to the reaction of 2-sulfinyl acrylates, **32b** and **32c**, with furan [46, 49]. This could result from the fact that the formation of the adducts is not observed under the thermal conditions. The addition of different catalysts (ZnI_2, $TiCl_4$, etc), did not solve the problem but afforded complex mixtures of reaction. This behavior could not be attributed to the lack of reactivity of the dienophile, because the less reactive 3-sulfinylacrylates are slowly transformed into their

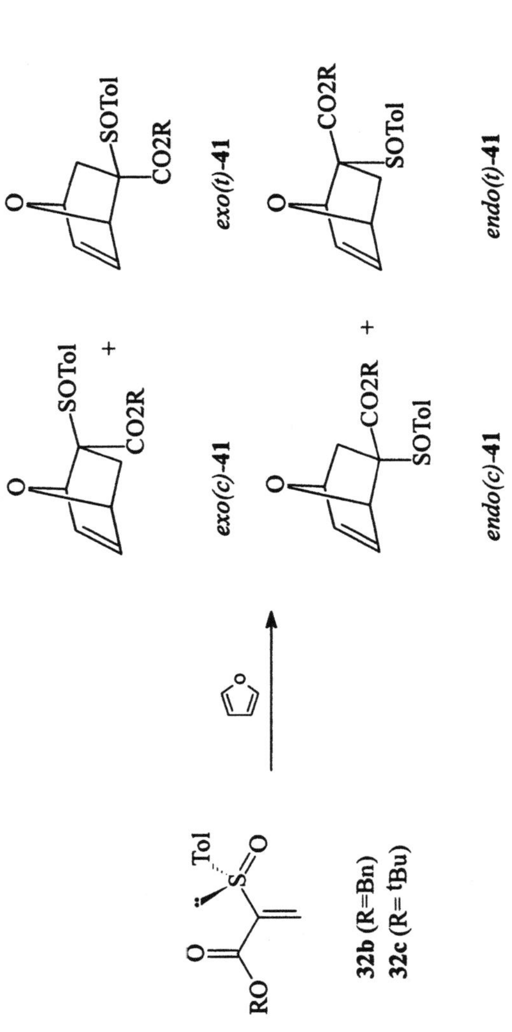

	P (Kbar)	t(h)	Conversion	exo(c)-41	exo (t)-41	endo(c)-41	endo(t)-41
32b	13	24	95%	24	44	11	21
32c	4	120	65%	29	59	4	8
	8	6	90%	22	45	15	18
	13	24	95%	22	45*	-	-

* bis-adduct were also obtained

Scheme 19

corresponding adducts under these conditions. Therefore, we assume that the retro-Diels-Alder reaction of the adducts obtained from 2-sulfinylacrylates must be easier than that of those obtained from 3-sulfinyl derivatives. This assumption prompted us to investigate the effect of high pressures, which strongly favour Diels Alder reactions (bimolecular process) but retard retro-Diels-Alder (dissociative unimolecular process). Reaction of **32b** with furan at **13** kbar yielded a mixture of the four possible adducts **41** in significant amounts (Scheme 19). Separation and characterization of these compounds were difficult because they spontaneously transformed into the starting materials at atmospheric pressure. The stability of the adducts obtained from **32c** was higher, and therefore most of the studies were performed with this dienophile.

The major adducts obtained in reactions of **32c** with furan have been used to prepare both (+)-shikimic and (+)-5-*epi*-shikimic acids in their non-natural configurations, following the sequence depicted in Scheme 20 [49].

The only report of 2-sulfinyl butenolides appeared in 1993 [50]. The synthesis of 5-ethoxy-3-*p*-tolylsulfinyl-2(5*H*)-furanones (**42a** and **42b**) and their behavior as dienophiles in asymmetric Diels-Alder reactions with cyclopentadiene were studied. This paper evaluates the relative ability of the two chiral centers at the dienophile (sulfur and C-5) to control the stereochemical course of the reaction. From the results obtained it was concluded that both chiral centers have a

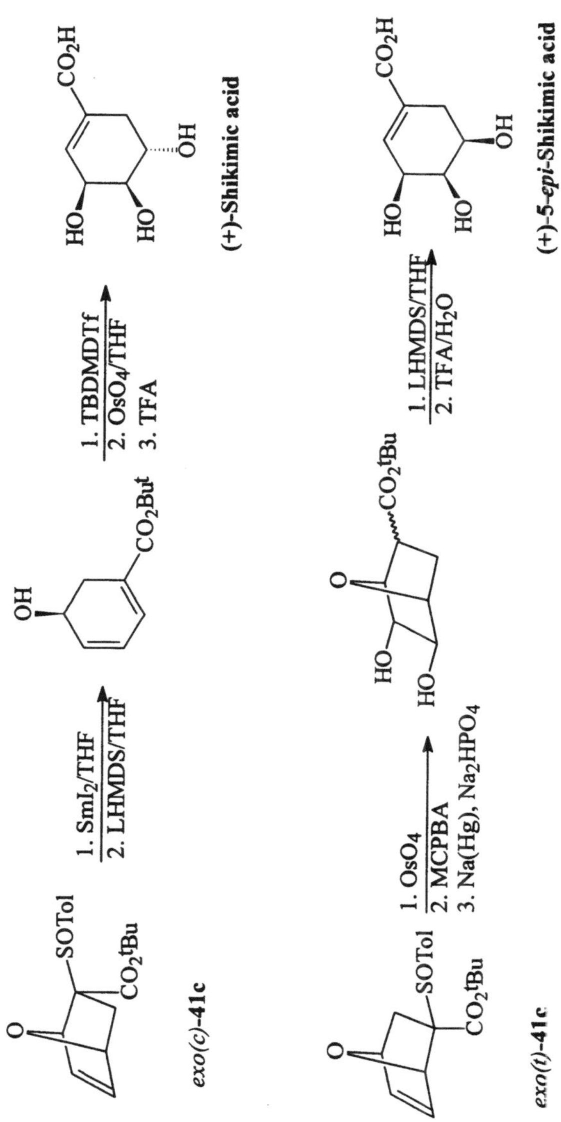

Scheme 20

similar influence, prevailing control by C-5 for uncatalyzed reactions and control by the sulfoxide in reactions performed under $ZnBr_2$ catalysis. Highly stereoselective transformations were observed only under conditions favoring the rotamer around the C-S bond in which the lone pair of electrons at sulfur is oriented toward the face containing the hydrogen at C-5. Hence, reactions of **42a** catalyzed by $ZnBr_2$ (which stabilizes the chelated *s-trans* conformation with the bottom face doubly favored), and those of **42b** under thermal conditions (favoring the *s-cis* rotamer on electrostatic grounds, with the upper face more accessible) took place with almost complete π-facial selectivity. They afford a mixture of *endo* and *exo* adducts **43** (Scheme 21). Taking into account that cycloadditions of chiral 5-alkoxybutenolides lacking the sulfinyl group occur also with a complete control of the stereoselectivity [51], the incorporation of a sulfinyl group in these substrates did not provide any important advantage from a synthetic perspective.

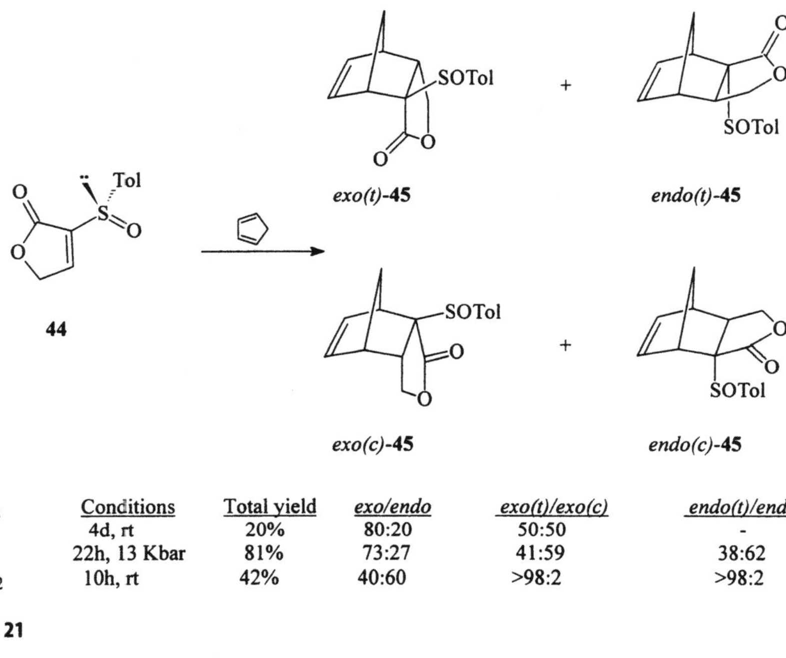

Catal.	Conditions	Total yield	exo/endo	exo(t)/exo(c)	endo(t)/endo(c)
-	4d, rt	20%	80:20	50:50	-
-	22h, 13 Kbar	81%	73:27	41:59	38:62
$EtAlCl_2$	10h, rt	42%	40:60	>98:2	>98:2

Scheme 21

The reactivity of the 3-*p*-tolylsulfinyl butenolide **44** has also been investigated [52]. The main results obtained in its reactions with cyclopentadiene are depicted in Scheme 22. Good yields were obtained only under $EtAlCl_2$ catalysis or high pressures, revealing a low dienophilic reactivity. The stereochemical course of the reactions of **44** with cyclopentadiene under different conditions was almost identical to that observed with 2-*p*-tolylsulfinyl-2-cyclopentenones (see below). Additionally, a comparative study of the results in Schemes 22 and 16 (these last obtained from acrylates **32**) shows that cyclic and acyclic dienophiles exhibit complete π-facial selectivity in the presence of chelating agents ($EtAlCl_2$ for **44** and ZnX_2 for **32**). It suggests an almost identical structure of the chelated

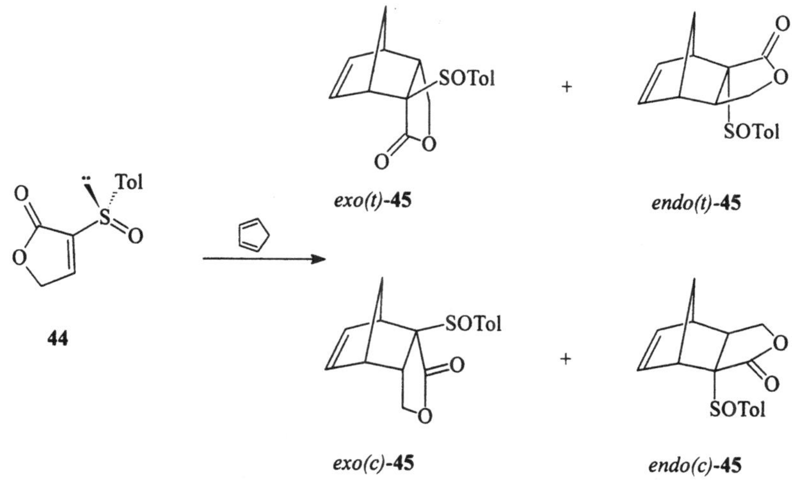

Catal.	Conditions	Total yield	*exo/endo*	*exo(t)/exo(c)*	*endo(t)/endo(c)*
-	4d, rt	20%	80:20	50:50	-
-	22h, 13 Kbar	81%	73:27	41:59	38:62
$EtAlCl_2$	10h, rt	42%	40:60	>98:2	>98:2

Scheme 22

species reacting with diene (*s-trans* rotamer around the C-S bond, B in Fig. 3), which has diastereotopic faces well-differentiated from a steric point of view. There are some differences in the *endo/exo* selectivity, which are presumably due to the influence of the alkyl substituent in **44** (which is absent in **32**) increasing the amount of *endo*(*t*)-**45**. In the absence of catalyst, the behavior of butenolide and acrylates are quite different. Compound **44** is much less reactive (it requires high pressure to obtain good yields) and exhibits a lower π-facial selectivity than acyclic acrylate **32**. The significant contribution of **A** and **A'** (Fig. 4) as the reactive rotamers in butenolide **44** (see later – sulfinyl quinones), and the main contribution of conformations like **B** in acrylates (nonexistent for butenolides), were invoked to explain differences in the π-facial selectivity (Fig. 4).

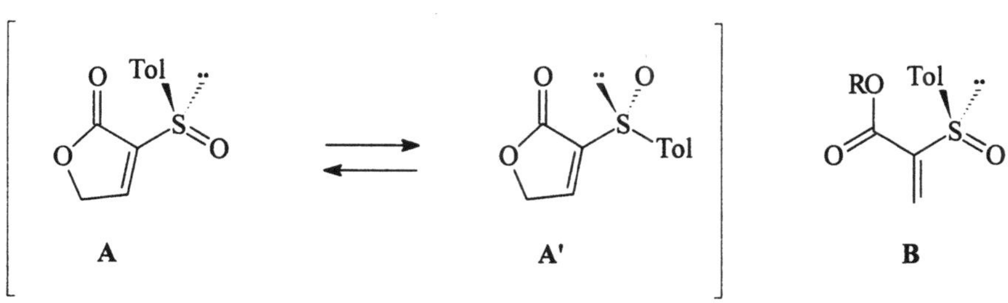

Fig. 4

Reactions of **44** with acyclic dienes (piperylene and 1-vinylcyclohexene) were unsuccessful under different thermal and catalytic conditions. Reactions were therefore conducted under high pressures without catalyst, affording mixtures of desulfinylated adducts (low regioselective desulfinylation) with rather low optical purity (ee = 45%). Bearing in mind the almost complete *exo*-selectivity of reactions with acyclic dienes, the above results were explained by assuming a moderate π-facial selectivity for these cycloadditions. These assumptions were supported by the results obtained in reactions of **44** with the Dane's diene under catalyzed ($EtAlCl_2$, 1 equiv.) and non-catalyzed (high pressure) conditions [52]. These reactions gave mixtures of only two adducts, *exo*(*t*)-**46** and *exo*(*c*)-**46** (Scheme 23), thus confirming the total *exo*-selectivity but moderate π-facial selectivity of the acyclic dienes. The latter could be inverted by using 2 equiv. of $EtAlCl_2$ as Lewis acid. In this sense, the behavior of **44** was almost identical to that observed when using 2-*p*-tolylsulfinyl cyclopentenone as starting material (see later for more details). The regioselectivity of these cycloadditions was also complete.

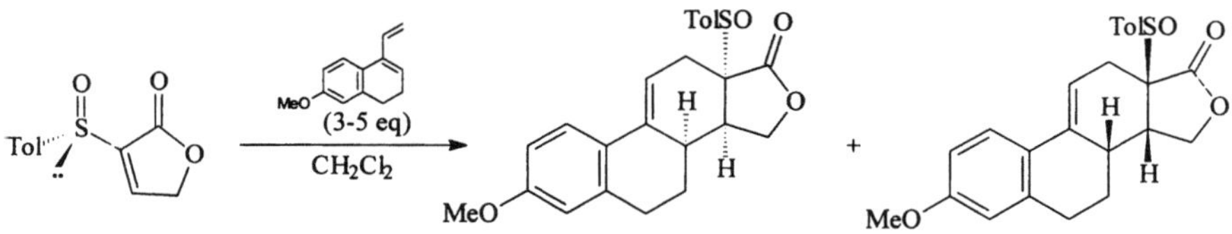

44 *exo(t)*-**46** *exo(c)*-**46**

Scheme 23

2.1.2.2
Carbonyl Derivatives

In contrast to the sulfinyl acrylates, the behavior of the enantiomerically pure sulfinyl enones as dienophiles has been very little studied. The first report in this field was due to Maignan et al. [53], who described the synthesis of several sulfinyl enones and the reaction of 3-*p*-tolylsulfinyl butenone **47** with cyclopentadiene. The reaction required 12 h at room temperature to reach completion, and a 60:40 mixture of the two *exo* adducts was obtained (Scheme 24). This result suggested that the *endo*-orientating character of the carbonyl group is much higher than that of the sulfinyl one, thus resulting in only *exo*-adducts (*endo* with respect to the carbonyl group). By contrast, the π-facial selectivity is very low.

In the original paper there is no discussion of the stereochemical course of this reaction. Nevertheless, the stereochemistry of the major adduct *exo*(*t*)-**48** suggested that the favored approach of the diene must take place toward the less hindered face of dienophile, with the sulfoxide adopting the spatial arrangement of conformation **47A′** (Scheme 24), which would be expected to exhibit a similar π-facial selectivity to that of the rotamers with the sulfinyl oxygen in an *s-trans* arrangement. This behavior is similar to that observed for **44** (see Fig. 4)

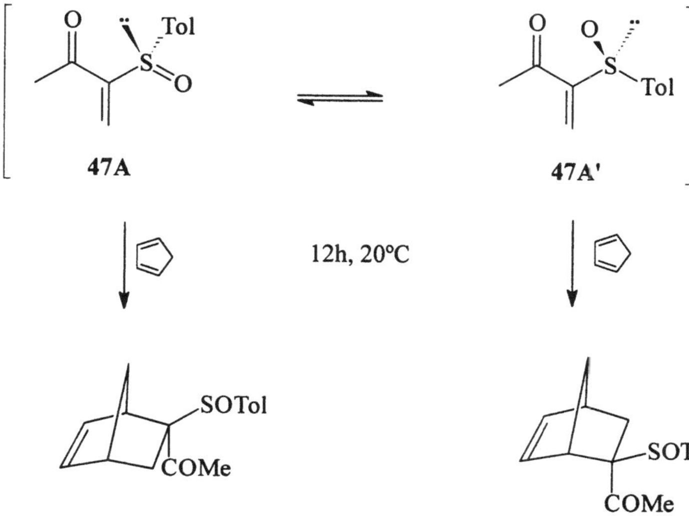

Scheme 24 *exo(c)*-**48** (minor) *exo(t)*-**48**(major)

and strongly contrasts with the preferred adaptation of the *s-cis* conformation in 2-*p*-tolylsulfinyl acrylates, where *endo*(*c*) and *exo*(*c*) adducts were isolated as the major products (see above). Recently, after the synthesis of several optically pure (*E*)-sulfinyl enones [54], their behavior with cyclopentadiene in the presence of different Lewis acids was evaluated [55]. π-Facial selectivity was moderate or low in all cases (< 50%), whereas the low *endo/exo* selectivity observed under thermal conditions (ca. 1 : 1 mixtures) was increased in the presence of Lewis acids, the best results being obtained in the presence of $BF_3 \cdot OEt_2$ (93:7 mixture).

Reactions of optically pure sulfinyl enones with acyclic dienes have never been reported, perhaps as a result of the discouraging results obtained from racemic substrates [56]. The reaction of α-phenylsulfinyl α,β-unsaturated ketones with butadiene and 2,3-dimethyl butadiene, under Lewis acid catalysis, yielded the cyclohexadienes resulting from spontaneous elimination of sulfenic acid (precluding any conclusion about the stereoselectivity of the reaction). Analogous results were found by Nishio et al. [57] from reactions of β-phenylsulfinyl α,β-unsaturated ketones with acyclic dienes. The bicyclic adducts derived from cyclopentadiene are more stable [57]. In this case, the reactions gave only the *exo*-sulfinyl adduct, but these stereochemical results were neither confirmed nor discussed in the paper.

The use of sulfinyl cycloalkanones **49a** and **49b** as dienophiles was first reported in 1989 [58]. Their reactions with cyclopentadiene were possible only in the presence of aluminum catalysts. The use of 1.2 equiv. of $EtAlCl_2$ (which proved to be the most efficient catalyst) afforded a mixture of only two adducts *endo*(*t*)-**50** and *exo*(*t*)-**50**, the major ones being the *endo* adducts, and the stereoselectivity being higher for the six-membered ring (Scheme 25). The complete π-facial selectivity observed in these reactions was explained by as-

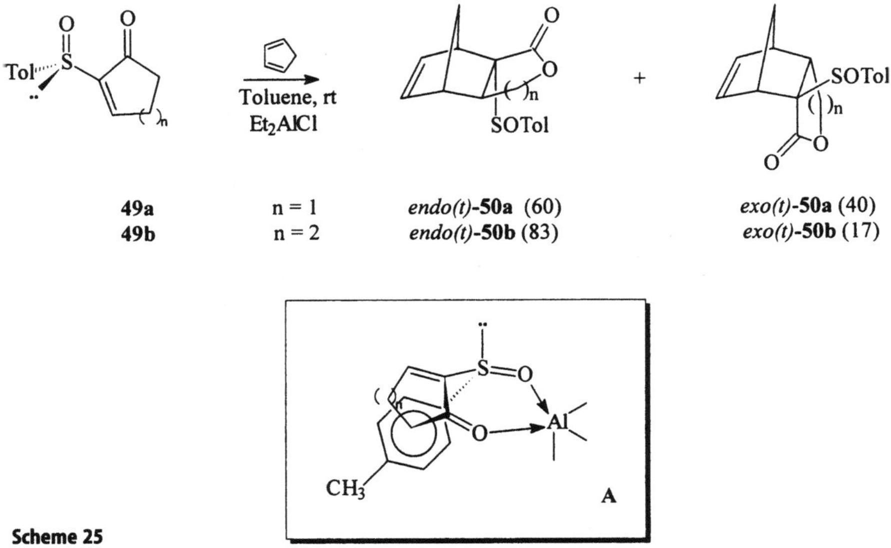

Scheme 25

suming the formation of the chelated species A (Scheme 25) in the presence of the catalyst, which has only one face accessible to the attack by the diene.

2-*p*-Tolylsulfinylcyclopentanone **49a** was further used in the enantioselective synthesis of steroid skeletons [59]. Its reaction with Dane's diene in the presence of $EtAlCl_2$ (1 equiv.) afforded only one adduct (*exo*(*t*)-**51**) with complete regioselectivity (controlled by the aromatic C-2 substituent) and *endo/exo* selectivity (the *endo*-orientating character of the keto group is predominant). The use of 2 equiv. of catalyst inverts the π-facial selectivity, yielding a mixture of the two possible *exo*-adducts, the major one being *exo*(*c*)-**51** (Scheme 26). The chemical instability of the adducts precluded the unequivocal assignment of their absolute configurations. Assignment was therefore based on the assumption that in the presence of 1 equiv. of catalyst, the approach of diene taking place from the less hindered face of the chelated species A (Scheme 25). The exclusive formation of the *exo*-**51** adducts with Dane's diene contrasts with the predominance of the *endo* adducts obtained with cyclic dienes (Scheme 25). This increase of the *exo* selectivity (with respect to the sulfinyl group) observed in reactions with acyclic dienes has also been found for sulfinyl butenolides (see above) and other sulfinyl esters such as acrylates and maleates (see later). The use of two equiv. of the catalyst determines the reversal π-facial selectivity (*exo*(*c*)-**51** is now favored). It was explained by assuming the formation of the species B (Scheme 26), involving a double association of **49a**, which must favor the *s-cis* arrangement of the sulfinyl oxygen, and therefore exhibit the opposite π-facial selectivity. The study of the reactions of **49a** and **49b** with cyclopentadiene in the presence of 2 equiv. of $EtAlCl_2$ would allow clarification of the above assumption.

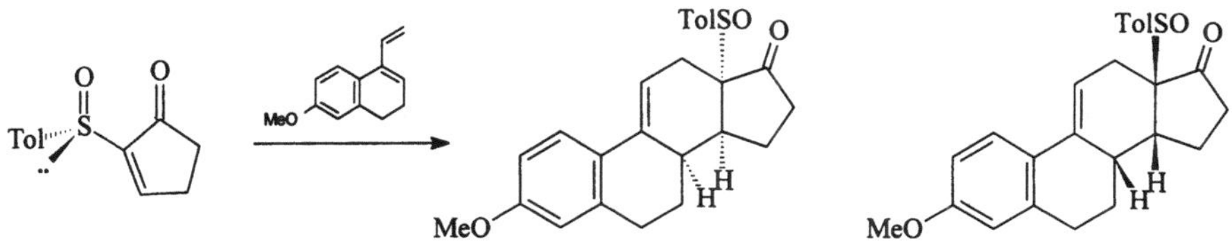

49a	catal.	*exo(t)*-51	*exo(c)*-51
	$EtAlCl_2$ (1 eq)	98	2
	$EtAlCl_2$ (2 eq)	29	71

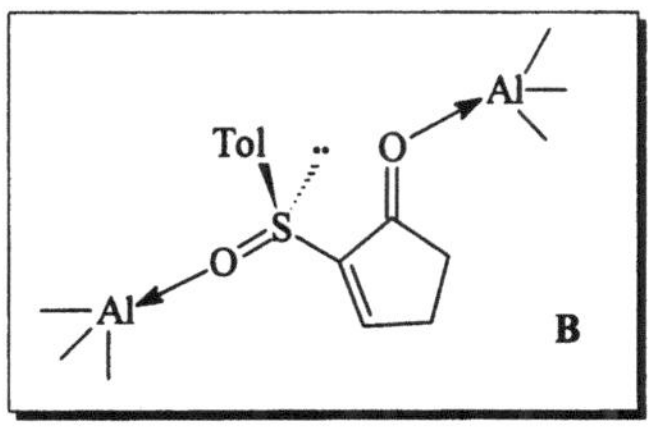

Scheme 26

2.1.2.3
Amides and Nitriles

Very few papers on sulfinyl amides have been published. In 1981, Danishefsky reported the use of compound **52**, with *β*-sulfinyl acrylamide structure, in the total synthesis of pretyrosine [60, 61]. The sulfoxide **52** was obtained as a mixture of diastereoisomers by MCPBA oxidation of the corresponding thioether, which in turn had been prepared by using pyroglutamate derivatives as source of chirality. Reaction of **52** with Danishefsky's diene was carried out in benzene under reflux for 22.5 h (suggesting a moderate reactivity). The reaction product (presumably a mixture of adducts which was not studied, thus precluding any stereochemical information about the Diels-Alder reaction) was transformed into compound **53** by treatment with 2.5% acetic acid (Scheme 27). As we can see, in these papers, the presence of the sulfinyl group is used only to take advantage of its ability as a precursor of the double bond, its role in the stereoelectronic course of the cycloadditions being irrelevant.

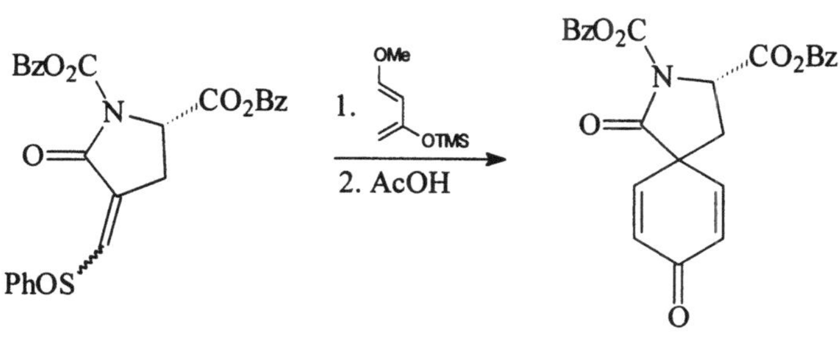

Scheme 27 **52** **53**

The synthesis of the acyclic *N,N*-dimethyl, 2-*p*-tolylsulfinyl acrylamide and its use as a dienophile was reported in 1990 [62]. A rather low reactivity toward cyclopentadiene (long reaction times under catalytic conditions were required to obtain moderate yields) and the lack of diastereoselectivity (a mixture was formed of the four possible adducts in significant amounts) were the most relevant findings from the use of this dienophile.

In 1978, Weiler and Brennan [63] reported the use of racemic 4-*iso*-thiazolin-3-one-1-oxide (**54**) as a dienophile, the structure of which could be considered to be a cyclic sulfinyl acrylamide. It undergoes facile cycloaddition (temperatures under 60 °C are required) with cyclopentadiene, 2,3-dimethyl-1,3-butadiene and 1,3-cyclohexadiene, to afford only one product in each case (Scheme 28). Reactions with anthracene and hexachloro cyclopentadiene required temperatures above 100 °C, and were effectively catalyzed by $AlCl_3$. Although the stereochemistry of the obtained adducts was not ascertained, the authors suggest that the major one is the result of an *endo*-approach, which must be favored on the basis of mechanistic considerations.

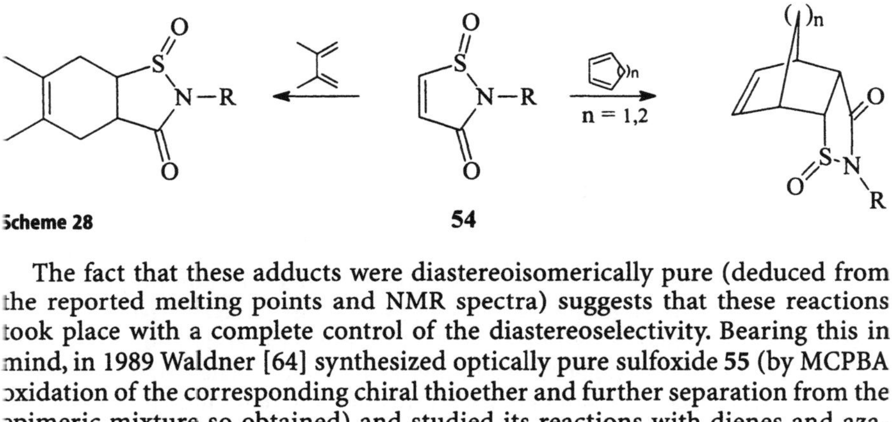

Scheme 28

54

The fact that these adducts were diastereoisomerically pure (deduced from the reported melting points and NMR spectra) suggests that these reactions took place with a complete control of the diastereoselectivity. Bearing this in mind, in 1989 Waldner [64] synthesized optically pure sulfoxide **55** (by MCPBA oxidation of the corresponding chiral thioether and further separation from the epimeric mixture so obtained) and studied its reactions with dienes and azadienes (Scheme 29). A high π-facial selectivity (almost complete with cyclopentadiene and azadienes) was observed in all cases, the approach of the diene from the dienophilic face supporting the lone pair of electrons at sulfur being favored. In the case of the reaction with cyclopentadiene, the *endo*-selectivity was complete as well.

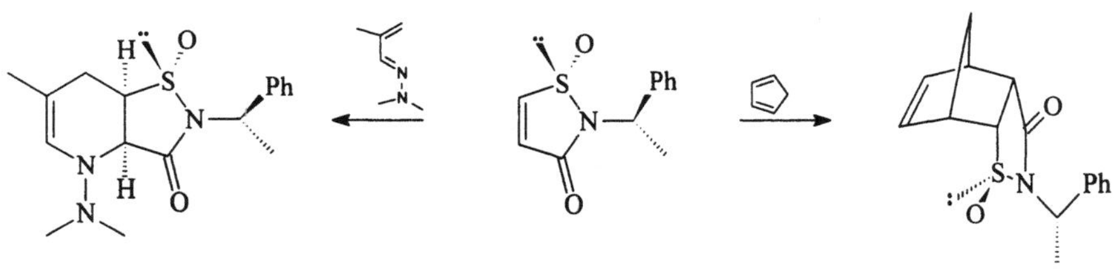

(>98% *de*) **55** (96% *de*)

Scheme 29

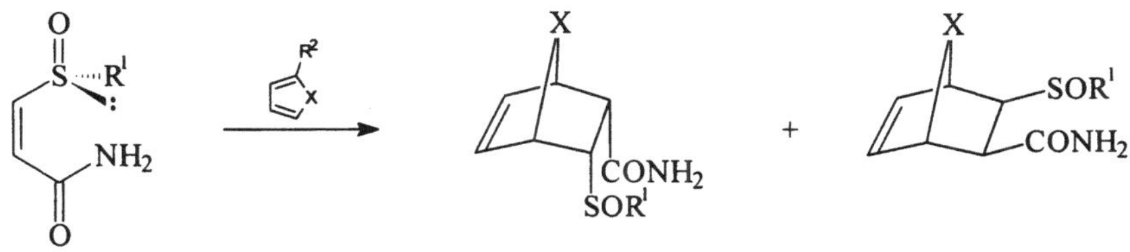

56 *endo(t)*- **57** *exo(t)*- **57**

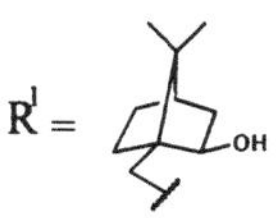

X	R^2	P		
CH_2	H	1 atm	87	5
CH_2	H	12 Kbar	81	9
O	H	12 Kbar	81	-
O	OMe	12 Kbar	92	8

Scheme 30

The most recent report in this field is related to the use of optically pure 3-(2-*exo*-hydroxy-10-bornyl)propenamide **56** [43]. This compound reacted with cyclopentadiene under both atmospheric and high pressure conditions in CH_2Cl_2/methanol to give similar mixtures of *endo* and *exo* cycloadducts (*de* ~ 80 – 88%). Reactions with furan and 2-methoxyfuran took place only under high pressure (12.6 kbar). With furan, the *endo*-adduct was obtained as a single diastereoisomer (82% yield), whereas with 2-methoxyfuran a 92:8 mixture of *endo*-**57** and *exo*-**57** adducts was obtained (Scheme 30).

The dienophilic behavior of sulfinyl acrylonitriles has been recently studied. The most interesting results have been obtained from Z-isomers **58**, which were obtained by stereoselective hydrocyanation of the corresponding alkynyl sulfoxides [65]. Compound **58a** reacted with cyclopentadiene under thermal conditions (reflux in CH_2Cl_2, 1 h) yielding a >90:<10 mixture of *endo*(*t*)-**59a** and *exo*(*t*)-**59a** adducts (Scheme 31), evidencing a complete control of the π-facial selectivity. In the presence of $ZnBr_2$, the reactivity is slightly higher (1 h at room temperature) but the *endo/exo* ratio and the π-facial selectivity remain unaltered. Compounds **58b** (R = *n*-Bu) and **58c** (R = CH_2Ph) exhibited a lower reactivity than **58a** (longer reaction times were required) to afford mixtures of *endo*(*t*)-**59b** + *exo*(*t*)-**59b** and *endo*(*t*)-**59c** + *exo*(*t*)-**59c**, respectively, under thermal or catalyzed ($ZnBr_2$) conditions (Scheme 31). The presence of the alkyl groups at dienophile decreased the reactivity and the *endo* selectivity of the reaction, but it does not affect the π-facial selectivity, which is completely controlled by the sulfinyl group. Both adducts are formed by attack of the diene from the only accessible face of the dienophile adopting the *s-trans* conformation.

The complete π-facial selectivity observed for these reactions, which is probably the highest so far reported with acyclic sulfinyl dienophiles, can be rationalized by assuming that conformational equilibrium around the C-S bond was completely restricted as a consequence of an important dipolar repulsion between cyano and sulfinyl groups. It determines that the latter one adopts the *s*-

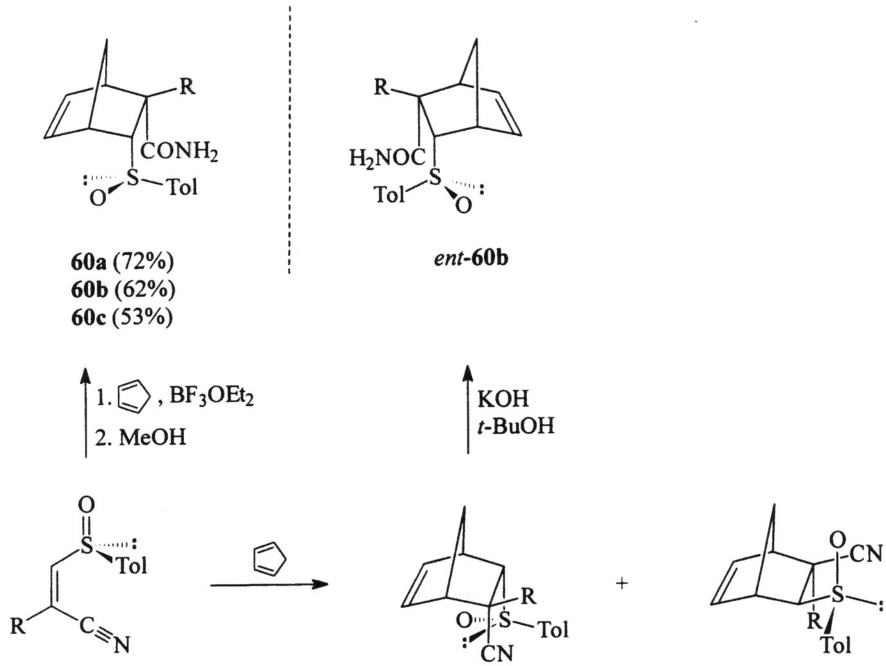

60a (72%)
60b (62%)
60c (53%)

ent-**60b**

58 *endo(t)*- **59** *exo(t)*- **59**

dienophile	catalyst	*endo(t)*-**59**	*exo(t)*-**59**
58a (R = H)	-	>90	<10
	$ZnBr_2$	>90	<10
58b (R = *n*-Bu)	-	48	52
	$ZnBr_2$	45	55
58c (R = CH_2Ph)	-	45	55
	$ZnBr_2$	58	42

Scheme 31

trans arrangement shown in Scheme 31, which exhibits a marked steric differentiation between the diastereotopic faces. This arrangement cannot be modified by $ZnBr_2$ because chelated species involving CN group are not possible.

One of the most interesting findings of this study was the behavior of dienophiles **58** in the presence of BF_3. When cycloadditions were catalyzed by this Lewis acid (which is more efficient than $ZnBr_2$) and MeOH was added prior to the isolation of the resulting products, compounds **58a–c** gave adducts **60a–c**, respectively (Scheme 31). As we can see, the endo-selectivity of these reactions is complete (even in the case of dienophiles **58b** and **58c**), as well as their π-facial selectivity, but this is opposite to that observed under thermal conditions (Scheme 31). Moreover, amides are obtained instead of nitriles, thus indicating that hydrolysis of the CN group has taken place. Finally, the configuration of the sulfinyl group at the adducts **60** is the opposite to that of **59**. This latter point was established by chemical correlation. The hydrolysis (KOH/*t*-BuOH) of the CN group of compound *endo*(*t*)-**59b**, yielded the amide *ent*-**60b**, the enantiomer of **60b** exhibiting the same spectroscopic parameters and identical specific rotation but with the opposite sign.

The stereochemical course indicated in Scheme 32 was suggested to explain the results obtained in the presence of BF_3. The activation of the CN group by the catalyst would allow it to be intramolecularly attacked by the nucleophilic sulfinyl oxygen, forming the cyclic intermediate species A, with the sulfinyl oxygen in *s-cis* arrangement (instead the *s-trans* favored under thermal conditions, Scheme 31). It justifies the π-facial selectivity of the cycloaddition was reversed. The *endo*-orientating character of the heteroatomic functionality in **A** must be quite high, being able to attain to a complete control of the *endo*-selectivity even in the case of **58b** and **58c**. The resulting adducts **B** would be transformed into sulfinyl amides **60** by attack of methanol. This attack must produce the total inversion of the configuration at the sulfur atom (the sulfurane intermediate **C** depicted in Scheme 32 is suggested to explain such an inversion).

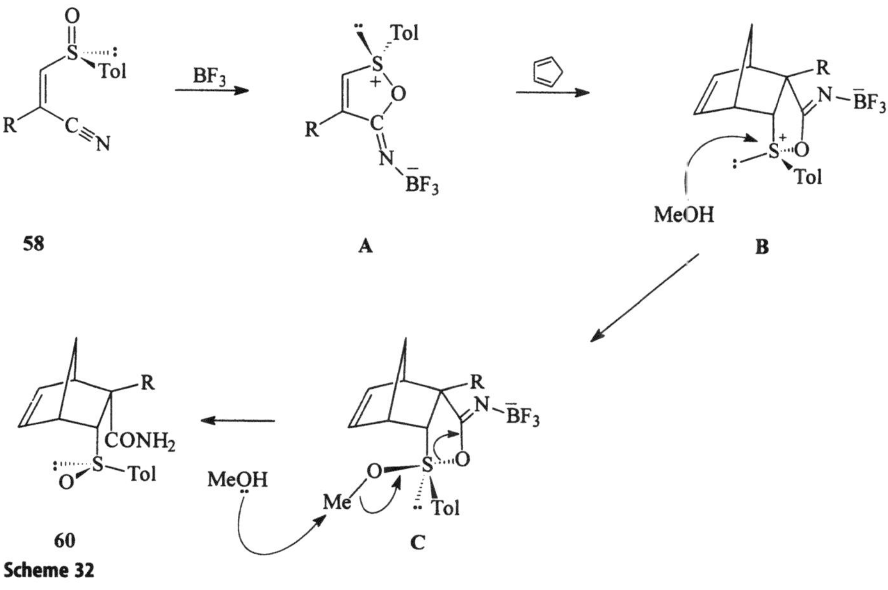

58 **60**

Scheme 32

Compound **58a** is able to react with furan (unpublished results) to yield a mixture of three adducts easily separated by chromatography, in the isolated yields indicated in Scheme 33 (13–23% of the starting dienophile is always recovered). Different conditions have been used, but the selectivity is very little dependent upon them. Furan must be used as solvent (2 days, room temperature), because in other cases (8 equiv.) the reaction does not work unless high pressures (4 kbar, 6 days) are used. The addition of $ZnBr_2$ or Me_2AlCl as catalysts reduces the reaction time (1 day, furan as solvent at room temperature). The stereochemistry of the major adduct *endo*-**61a** (51% isolated yield) was unequivocally established by X-ray diffraction studies. The *endo*/*exo* selectivity of the reaction ranges between 87/13 and 75/25, and the π-facial selectivity is >88%. The approach of diene from the lower face of dienophile in conformation depicted in Scheme 33 is clearly favored.

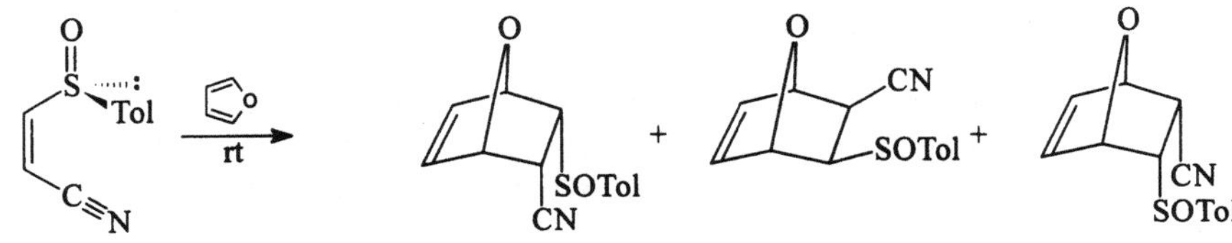

58a *endo*-**61a** *exo*-**61a** *endo*-**62a**

catalyst	reaction time (days)	*endo*-61a	*exo*-61a	*endo*-62a
-	2	51	11	2
$ZnBr_2$	1	43	16	4
Me_2AlCl	1	53	10	4

Scheme 33

2.1.2.4 Other Monoactivated α,β-Unsaturated Sulfoxides

The behavior of vinyl sulfoxides bearing other functional groups has been studied. Thus, reactions of cyclopentadiene with (*S*)-β-diethoxyphosphorylvinyl *p*-tolyl sulfoxide **63** [66] show that this dienophile exhibited a similar behavior to that of 2-sulfinyl acrylates **32**. Its moderated reactivity (10 days at room temperature in $CHCl_2$) but increased as the polarity of the solvent became higher (2 days at room temperature in H_2O/acetone), and in the presence of $ZnCl_2$ (1 day at –20°C in CH_2Cl_2) or BF_3OEt_2 (1 h at –20°C in CH_2Cl_2) as catalysts. High π-facial selectivity was observed only in reactions catalyzed by $ZnCl_2$, but the *endo*/*exo* selectivity is clearly poor (only a 75:25 mixture of *endo*(*t*)-**64** and *exo*(*t*)-**64** adducts were isolated). These results, quite similar to those obtained from 2-sulfinyl acrylates, were rationalized by assuming the formation of a chelated species involving the phosphoryl and sulfinyl oxygens of **63**, which must adopt the *s-trans* arrangement shown in Scheme 34.

The sulfonyl group has also been incorporated into the vinyl sulfoxide moiety in order to increase both its reactivity and selectivity. Thus, (*S*)-1-*t*-butylsulfonyl-1-*p*-tolylsulfonyl ethene **65** was used as a masked chiral ketene equivalent

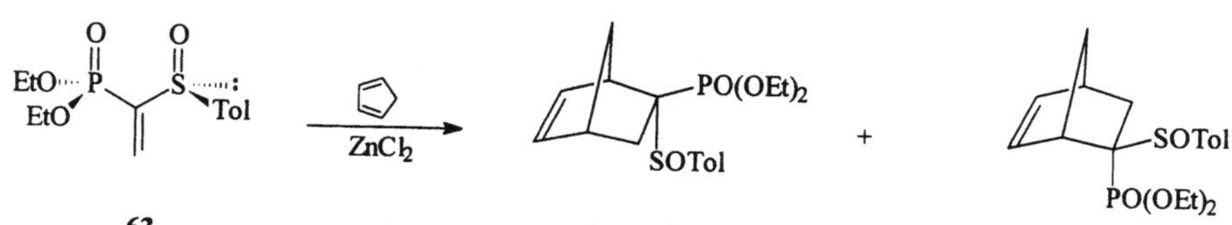

63 *endo(t)*- **64** (75) *exo(t)*- **64** (25)

Scheme 34

[67]. Its reactions with cyclopentadiene (Scheme 35) were satisfactory only in the presence of $ZnBr_2$, Eu(fod)$_3$, or SiO_2 as catalysts (stronger Lewis acids such as BF_3OEt_2, Et_2AlCl or $EtAlCl_2$ decomposed the dienophile), yielding ca. 90:10 mixtures of two adducts, from which the major one could easily be separated (ca. 60% yield). The chemical correlation of this adduct with (*R,R*)-dehydronorcamphor, and the assumption that the sulfinyl group of **65** adopts the *s-cis* conformation in order to avoid the electrostatic repulsion with the sulfonyl oxygens, allowed the authors to conclude that the isolated adduct had the *exo*(*c*)-**66** structure. The large magnitude of the steric interactions CH_2/SO_2Tol existing in the *endo* approach of the reagents (*exo* with respect to the sulfonyl group) also suggests such an assignment. The poor tendency of the sulfonyl group to become associated with Lewis acids would explain similar results being observed with chelating ($ZnBr_2$) and non-chelating (BF_3) catalysts. It is the opposite to that observed in the case of the 2-sulfinyl acrylates **32** and even in the case of the phosphorylsulfinyl ethylenes **63** (see above).

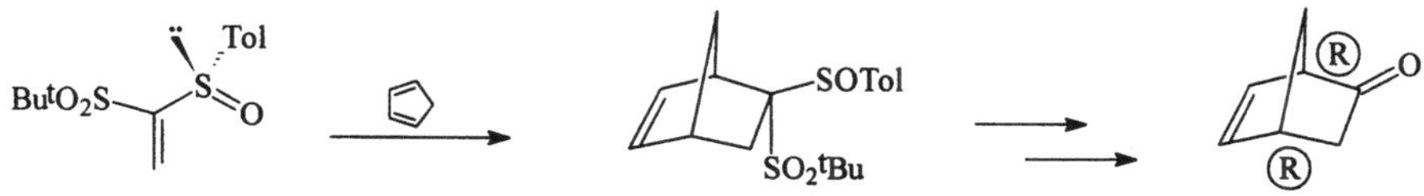

65 *exo(c)*-**66**

Scheme 35

(*E*)- and (*Z*)-2-arylsulfonylvinyl isobornyl sulfoxides **67** have also been studied [33]. Cycloaddition of the (*E*)-isomers afforded mixtures of diastereoisomers **68** as a consequence of a low stereoselective evolution of the dienophile (conformational mobility around the C-SO is not restricted). In contrast, (*Z*)-isomers react in a completely stereoselective manner, yielding the adducts resulting from *endo*-approach of diene to the less hindered face of the dienophile in its *s-trans* conformation (which exhibits the lowest electrostatic repulsion). The formation of one of the possible diastereoisomers is depicted in Scheme 36, but mixtures of epimers at sulfur were used as the starting material in the original paper.

The reactions of 1-thioderivatives of 1-trifluormethyl ethylenes have been recently reported [68]. The results obtained from racemic phenylsulfoxide (a mixture of the four possible adducts was isolated) indicate the small influ-

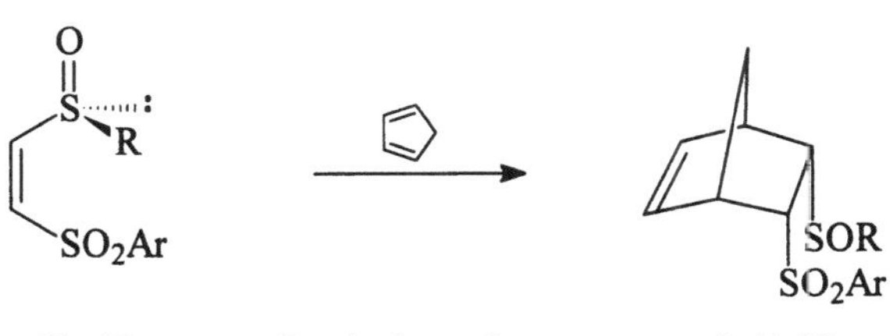

(*Z*)-**67** R = isobornyl *endo(t)*-**68**
Ar = Ph, *p*-ClC$_6$H$_4$

Scheme 36

ence of the CF_3 group on the stereoselectivity. This suggests that the use of these substrates in asymmetric cycloadditions would not be interesting.

2.1.2.5
Nitro Derivatives

The synthesis of the racemic (*E*)-*β*-phenylsulfinyl nitroethylene (**69**) and the study of its Diels-Alder reactions with cyclopentadiene and piperylene were reported in 1986 [69a]. The results obtained with the cyclic diene (Scheme 37) reported that the sulfinyl group exhibits a lower *endo* orientating character than the nitro group. The ability of the nitrogen function to control the regioselectivity with piperilene is also clearly the highest. The adducts resulting in reactions with acyclic dienes decompose into cyclohexadienes, which were easily oxidized to the corresponding benzenes (Scheme 37). Taking advantage of this behavior, the reaction of compound **69** with substituted 1-acetoxy-1,3-butadienes has been used to achieve the regioselective synthesis of *meta* and *para*-substituted nitrobenzenes [69b]. The synthesis of racemic (*Z*)-2-phenylsulfinyl 1-nitroalkenes and their use as synthetic equivalents of nitroacetylene in Diels-Alder reactions (itself too unstable to be used in these reactions) with acyclic dienes, was further reported [70].

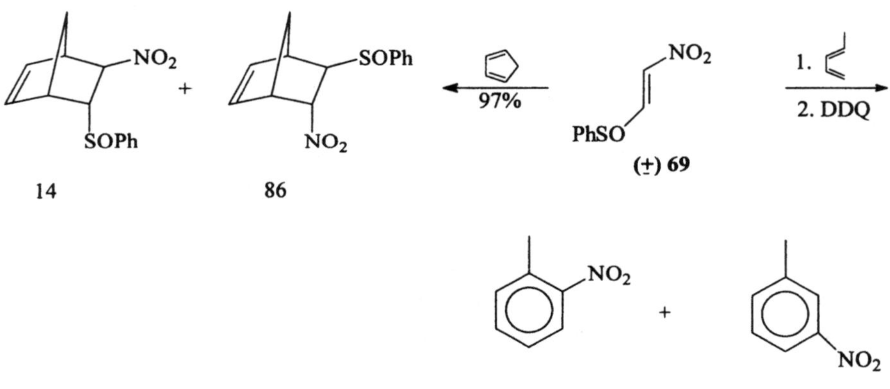

Scheme 37

The first papers dealing with the use of optically active sulfinyl nitroalkenes were published by Fuji et al. in 1991 [71, 72]. These papers describe the reaction of Danishefsky's diene with compound **70** (the dienophile containing a six-membered ring does not react) to afford a 1:1 mixture of two compounds (*endo*-**71** and *exo*-**71**) resulting from desulfinylation of the *endo*(*t*) and *exo*(*t*) adducts. The optical purity of these compounds (*ee* > 95%) indicates a complete π-facial selectivity in the cycloaddition. X-ray diffraction studies of dienophile **70** showed the *s-trans* conformation of the sulfinyl oxygen (like that depicted in Scheme 38) in the solid state. By assuming steric approach control for the Diels-Alder reaction, the stereochemistry of *endo*-**71** and *exo*-**71** suggests that such a conformation must also be preferred in solution (it will be the most stable one from an electrostatic point of view). This would explain that the approach of the diene had taken place toward the less hindered upper face of dienophile in Scheme 38 (that bearing the lone electron pair at sulfur).

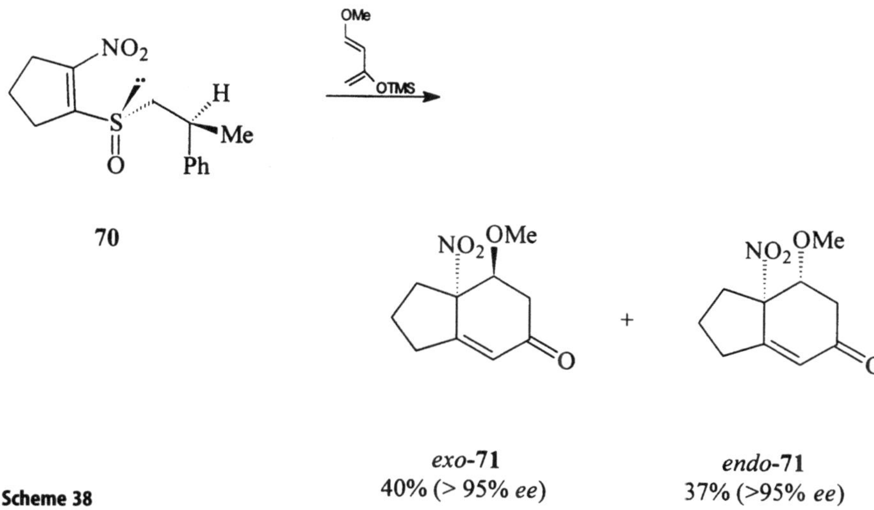

Scheme 38

As the reactivity of this dienophile with ordinary dienes, such as butadiene or cyclopentadiene, was very low, Fuji et al. have also investigated the effect of high pressures [73, 74] and Lewis acid catalysis [74] on [4+2] cycloadditions of some chiral sulfinyl nitroethylenes with non-activated conventional dienes. High pressures strongly increased the reactivity of cyclic dienophiles such as **70** but the stereochemical results are similar to those obtained at atmospheric pressure (complete regioselectivity and π-facial selectivity, but low *endo/exo*-selectivity). The lower temperatures required under high pressure precluded both the sulfinyl elimination and the aromatization processes. With acyclic sulfinyl nitroethylenes, the stereochemical results depend on the stereochemistry of the dienophile. Hence, (*Z*)-**72** reacts with 1,3-pentadiene at 8 kbar (the reaction does not occur at 1 atm) in a completely *endo* and π-facial selective manner, affording just one *endo*(*t*)-**73** adduct (Scheme 39), whereas its geometrical isomer (*E*)-**72** yielded a mixture of four different adducts, indicating moderate *endo/exo* and π-facial selectivities.

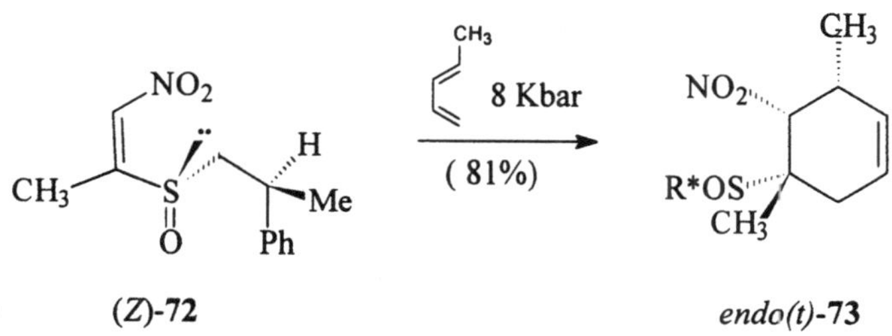

Scheme 39

Zinc halides effectively accelerated cycloadditions with cyclopentadiene, but they had little influence on the stereoselectivity. Thus, (*Z*)-**72** yielded only one adduct, *endo*(*t*)-**74**, whereas (*E*)-**72** gave a mixture of the four possible diastereoisomers (Scheme 40).

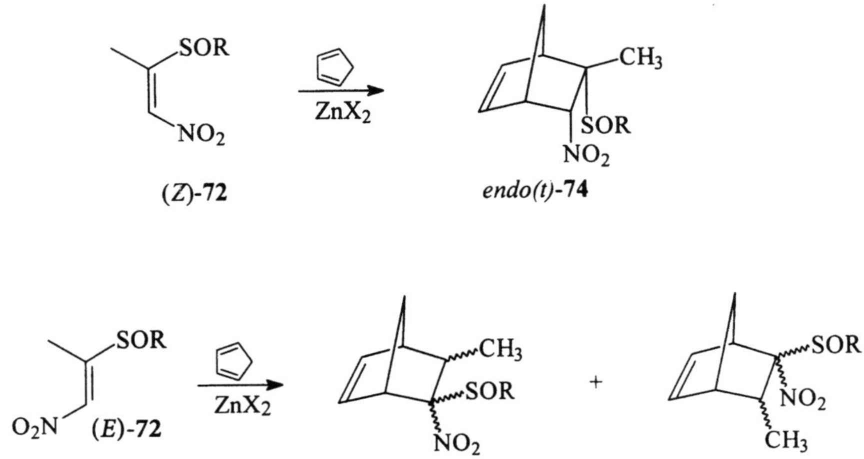

Scheme 40

The different stereochemical behavior of compounds (*E*)-**72** and (*Z*)-**72** was explained on the basis of their preferred conformations. In the absence of Lewis acids, the (*Z*)-olefin would adopt the *s-trans* arrangement of the sulfinyl group shown in Scheme 39 due to its strong dipole-dipole repulsive interactions with the oxygens at the nitro group (Fig. 5). The favored approach of the diene will then take place toward the less hindered face of dienophile – that containing the lone electron pair at sulfur – yielding *endo*-(*t*) adducts. The poor diastereoselectivity observed starting from the (*E*)-olefins was attributed to the lack of conformational rigidity around the C-S bond in solution.

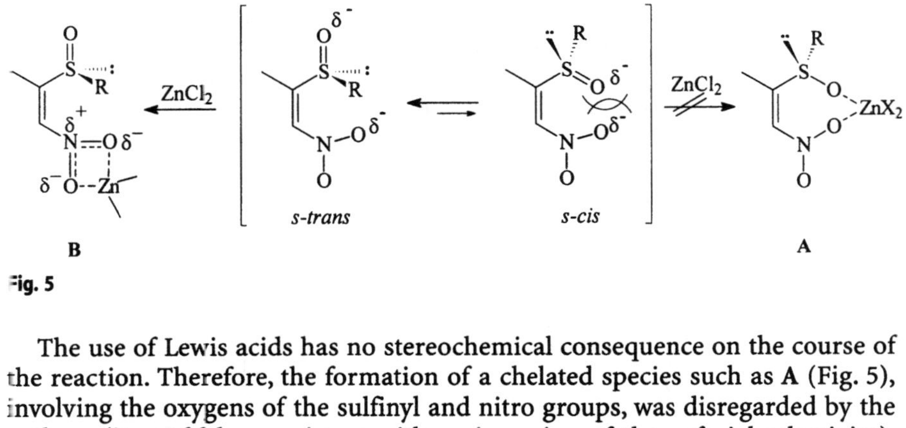

Fig. 5

The use of Lewis acids has no stereochemical consequence on the course of the reaction. Therefore, the formation of a chelated species such as **A** (Fig. 5), involving the oxygens of the sulfinyl and nitro groups, was disregarded by the authors (it would be consistent with an inversion of the π-facial selectivity).

They suggest another chelation, involving the two oxygens at the nitro group (**B** in Fig. 5) to rationalize the marked rate acceleration produced by addition of the Lewis acid, so maintaining the stereochemical preferences.

2.1.2.6
bis-Sulfinyl Ethylenes

Related to the use of masked ketenes in organic synthesis, the utility of ketene-dithioacetals [75], their racemic monoxides [76], and their tetroxides [77] has been profusely documented. Concerning the bis-oxides derived from these compounds, the synthesis of (*S,S*)-1,1-bis-(*p*-tolylsulfinyl)ethene (**75**) and their use as latent chiral ketene equivalents in Diels-Alder reactions was reported by Koizumi et al. in 1986 [78]. This dienophile, containing a C_2 axis, reacted with cyclopentadiene in a sealed tube at 60–70 °C, affording a 4:1 mixture of two adducts, **76a** and **76b**. These could not be separated and were further transformed into (+)-(1*R*)-5-norbornene (Scheme 41), the *ee* of which (54%) was consistent with the diastereoselectivity observed in the cycloaddition (60% *de*). This result allowed the establishment of the fact that the absolute configuration of the major adduct **76a** was that depicted in Scheme 41. Although the authors indicate that an extensive investigation about the scope and limitation of the dienophile **75** (including reactions with acyclic dienes as well as the use of Lewis acids as catalysts) was being made in their laboratory, these results have not yet published.

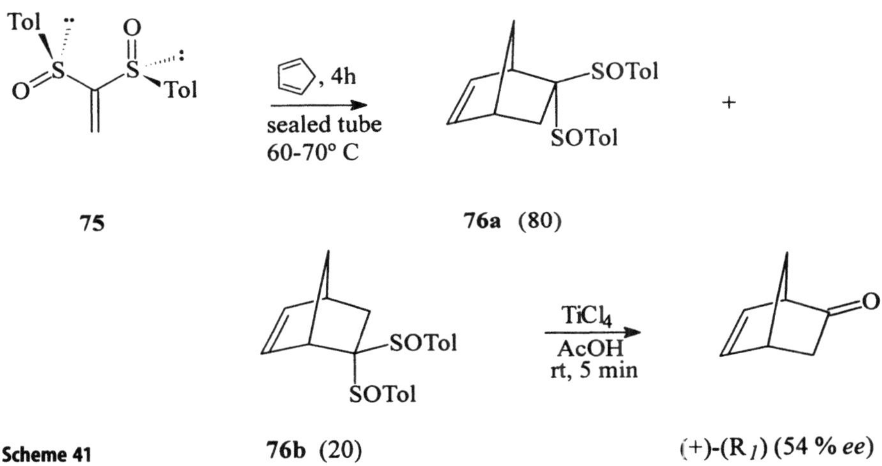

Scheme 41

When both sulfinyl groups were incorporated into a cyclic structure, the dienophilic reactivity was strongly improved, and the diastereoselectivity was dependent on the ring size. Thus, the first paper published dealt with the C_2-symmetric ketendithioacetal *S,S*-dioxide **77** (Scheme 42, obtained from 1,1′-binaphthalene-2,2′-dithiol), which reacts with cyclopentadiene at room temperature over 12 h (milder conditions than those required by **75**), yielding a 3:1 mixture of adducts [79]. The authors suggest a complete π-facial selectivity

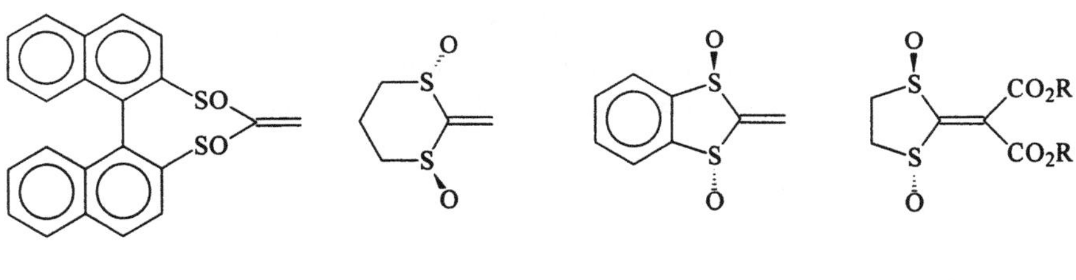

(±)-77 (±)-78 (±)-79 (±)-82

Scheme 42

(approach of the diene toward the opposite face to that bearing the sulfinyl oxygen being favored), but the reasons for the observed diastereoselectivity are not explained.

Reactivity and selectivity increased as the size of the ring containing the bis-sulfoxide moiety become smaller. Thus, racemic dienophile **78** [80] (Scheme 42) reacts with cyclopentadiene under BF_3 catalysis in 10 min at –78 °C, yielding a >25:1 mixture of adducts. In the absence of a catalyst, this dienophile is not able to react with 1-methoxybutadiene at room temperature, whereas this acyclic diene was completely transformed into a >25:1 mixture of adducts by reaction with compound (±)-**79** in 24 h [80]. Both dienophiles are able to react with furan under catalytic conditions. The authors conclude that dienophile **79** is in general more reactive and selective than is **78**.

The best results were obtained from dienophile **80**, which was synthesized in its optically pure form [81]. Reaction with cyclopentadiene occurred readily in propionitrile at room temperature, giving a 90:10 mixture of adducts **81a** and **81b** (Scheme 43). At –78 °C, adduct **81a** was obtained as a single diastereo-

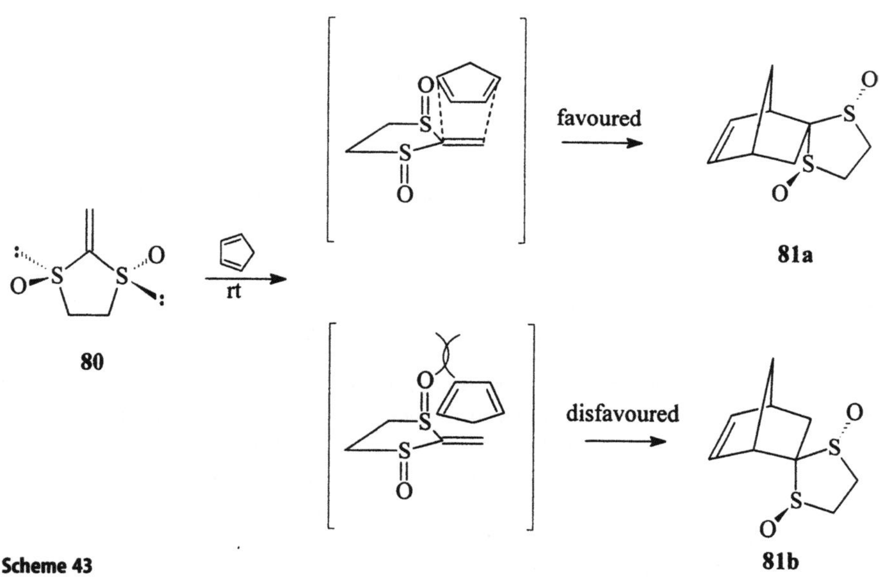

Scheme 43

isomer by using $BF_3 \cdot OEt_2$ as a catalyst. Acyclic dienes (without catalyst) and furan (under $SnCl_4$ catalysis) also reacted in a completely stereoselective manner under very mild conditions with dienophile **80**. This behavior, and the easy transformation of the bis-sulfoxide moiety into the carbonyl group (reduction and further hydrolysis of the resulting dithiolane), determine that dienophile **80** can be considered to be one of the best chiral ketene equivalents so far described.

The stereochemical outcome of the Diels-Alder reaction was rationalized from steric interactions existing in the two possible transition states (Scheme 43). The use of the Lewis acids as catalysts, which became associated with the sulfinyl oxygen thus increasing their effective size, improves the stereoselectivity of the reactions.

Racemic diesters **82** (Scheme 42), containing the bis-sulfoxide moiety incorporated into a five-membered ring, have also been used as dialkoxycarbonyl ketene equivalents [82]. They react with cyclopentadiene in 7–12 h at room temperature, affording mixtures of two adducts. The relative configuration of **82** was not unequivocally determined, but it was speculatively assigned on the basis of the conformational preferences of the precursor monosulfoxide and the presumably favored steric course of its MCPBA oxidation. We must point out the unexpectedly similar reactivity of dienophiles **80** and **82**, which is very difficult to explain given the increase in the dienophilic reactivity that two ester groups usually induce in ethylenic systems.

The latest paper in this field [83] concerns the synthesis of the (*S*,*S*)-1,1-bis-ethoxycarbonyl-2,2-bis-*p*-tolylsulfinylethene (**83**) and its dienophilic behavior. Reactions with cyclopentadiene occur neither under thermal conditions nor in the presence of Lewis acids, but required the use of high pressure (13 kbar) to afford an 87:13 mixture of **84a** and **84b** adducts. With acyclic 1-substituted dienes, the combined use of $ZnBr_2$ catalysis and high pressures was required to achieve high yields of cyclohexadienes **84c**, resulting from spontaneous pyrolytic desulfinylation of the adducts (Scheme 44). The optical purity of these cyclohexadienes (ee > 97%) revealed that both the regioselectivity and diastereoselectivity of these reactions are complete.

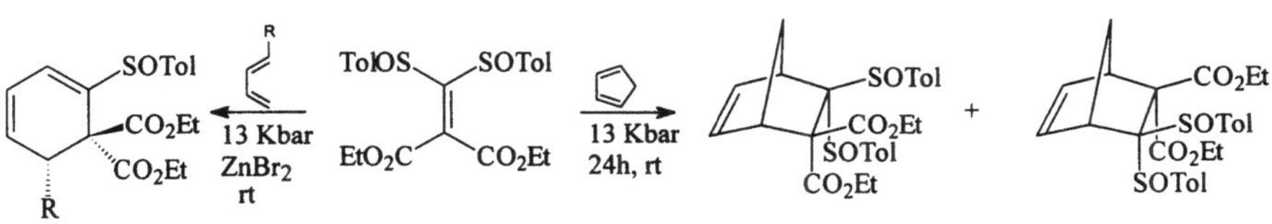

84c **83** **84a** (87) **84b** (13)

Scheme 44

The stereochemical course of these reactions and the configurational assignment of the adducts was rationalized by assuming that conformation **B** (*s-cis, s-trans*) must be favored over **A** (*s-trans, s-trans*), due to the strong electrostatic repulsion between the sulfinyl oxygens (Fig. 6). A complete π-facial selectivity (both *p*-tolyl groups are oriented toward the same face) and a higher *endo*-orientating character of the ester group coplanar with the dienophilic double bond, would be expected for the evolution of conformation **B**, which is in agreement with the observed experimental results.

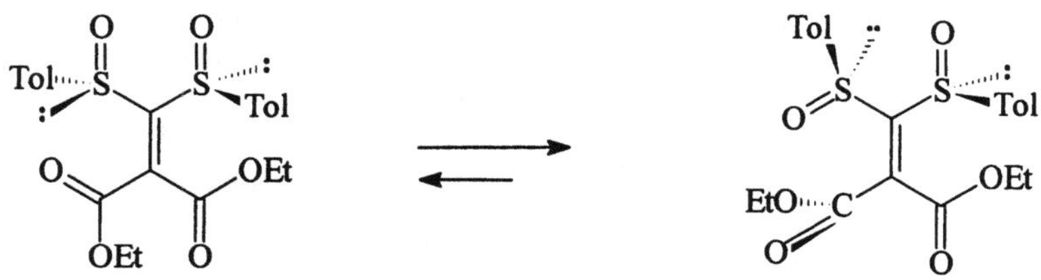

(s-*trans*, s-*trans*) (s- *cis*, s-*trans*)

Fig. 6 **A** **B**

One of the most interesting points that can be deduced from the studies carried out on the bis-sulfoxides is the much lower reactivity of acyclic dienophiles than that of cyclic examples (compare the results obtained from **80** and **75** or those from **82** and **83**). The influence of the conformation around the C-S bond (with higher restrictions in cyclic sulfoxides), on the dienophilic character of the double bond in vinyl sulfoxides, emerge as the most likely causes of the observed differences in reactivity (see later).

1,2-bis-sulfinyl ethylenes have been less studied [84]. The reaction of racemic (*Z*)-1,2-bis-phenylsulfinyl ethylene with cyclopentadiene (reported by Montanari et al. many years ago [84a]) only yields the *endo* adduct (12–24 h in refluxing benzene). In a similar way, enantiopure (–)-*trans*-benzo[d]-dithiine-*S*,*S*′-dioxide **86** reacts with cyclopentadiene and cyclohexadiene with complete *endo* selectivity [84b] (Scheme 45). The inclusion of the sulfinyl group in a cyclic rigid structure induces a substantial increase of the reactivity with respect to that of the acyclic bis-sulfoxides reported by Montanari et al. Furan is able to react, but the *endo*/*exo* selectivity is low, with the *exo* adduct being the major one.

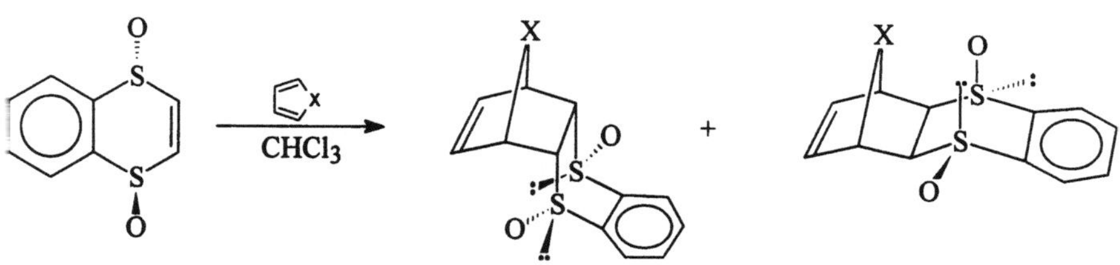

86 *endo* *exo*

X	T (°C)	t (h)	*endo*	*exo*
CH_2	60	0.5	100	-
$(CH_2)_2$	80	24	100	-
O	60	15	28	72

Scheme 45

2.1.3
Polyactivated Vinylsulfoxides

One of the main problems limiting the usefulness of monoactivated vinyl sulfoxides as chiral dienophiles derives from their moderate reactivity, and as a result most of the published papers report only their reactions with cyclopentadiene. In order to overcome this problem, other sulfinylethylenes containing two additional electron-withdrawing groups, further increasing the dienophilic reactivity, were studied. In this field, two main groups, of compounds have been investigated, where the sulfinyl group has been added to maleic and quinonic structures, presumably because of the high dienophilic reactivity of the pattern compounds.

2.1.3.1
Sulfinyl Maleates and Sulfinyl Maleimides

The first report of the use of racemic sulfinyl maleates as dienophiles appeared in 1983 [85]. This paper described the synthesis of 2-phenylsulfinyl maleate and fumarate, but only the reaction of cyclopentadiene with the first (±)-**87**, a synthetic equivalent of dimethyl acetylenedicarboxylate, was studied. The reaction takes place at room temperature on silica gel and the stereochemistry of the adducts remained undetermined because the isolated product is the compound resulting from the pyrolytic sulfinyl elimination, which occurs spontaneously under the reaction conditions used (Scheme 46).

In 1988 Koizumi et al. reported the first sulfoxide used as a chiral synthetic equivalent of dimethyl acetylene dicarboxylate [86a]. The synthesis of dimethyl $(R)_S$-2-(10-isobornylsulfinyl) maleate (**88**), its reaction with cyclopentadiene catalyzed by $ZnCl_2$, and the further transformation of the major adduct **89** into a half-ester **90** (which had been used as the starting material for the synthesis of carbocyclic nucleosides (–)-aristeromycin and (–)-neplanomicin) are described (Scheme 46).

Concerning the asymmetric Diels-Alder reaction, the highest diastereoselectivity was observed under $ZnCl_2$ catalysis at –20 °C. Under these conditions, the reaction provided exclusively one single *exo*-adduct, *exo*(*t*)-**89** (92%), along with a small amount of one *endo* isomer, *endo*(*t*)-**89** (6%). The stereochemistry of the former was unequivocally established from X-ray diffraction studies. This result showed the complete π-facial selectivity of the reaction and the predominance of the *endo* orientating character of the two ester groups with respect to the sulfinyl one. In this paper it was also reported that, in the absence of catalyst (room temperature, 12 h), the π-facial selectivity appears to be inverted, and a 93:7 mixture of *exo*(*c*)-**89** and *exo*(*t*)-**89** adducts (68%), along with only one *endo*(*c*)-**89** adduct (12%), were obtained (Scheme 46).

The latter reaction was further described by Koizumi et al. [86b] with slightly different results (lower facial selectivity for the *exo*-approach), in connection with the enantiodivergent synthesis of fused bycyclo[2,2,1]heptane lactones **91** (see Scheme 47). The key step of this transformation was the regioselective DIBAL reduction of only one of the two ester groups in the adducts, followed by

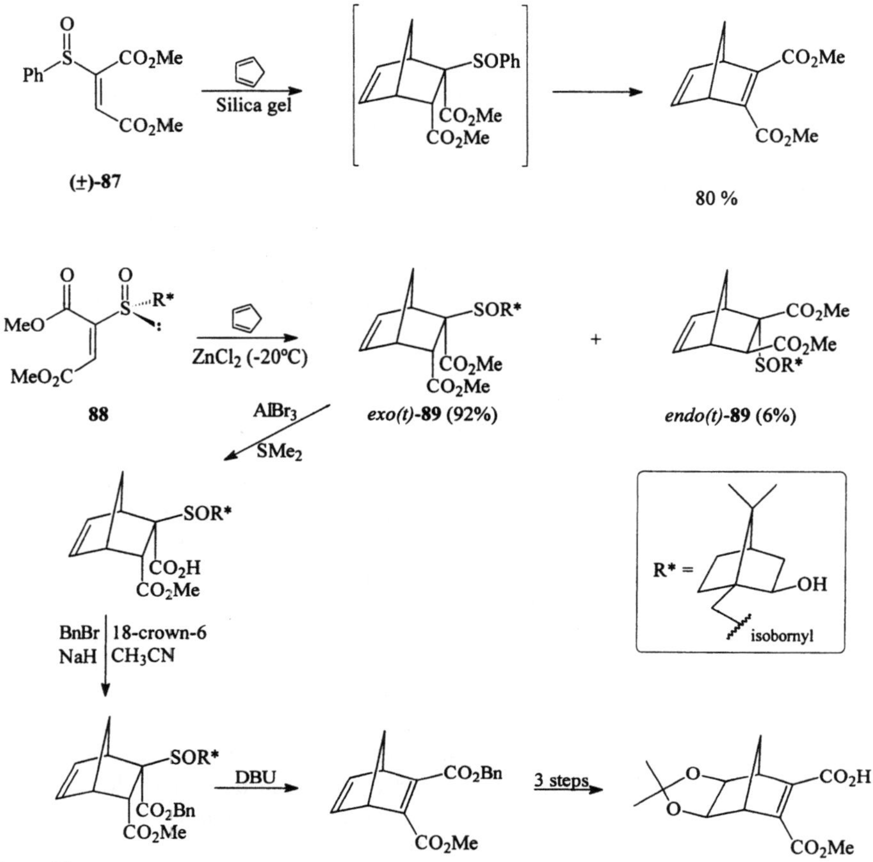

Scheme 46

reductive elimination of the sulfinyl group with SmI_2. The unsaturated lactones so obtained were easily transformed into (+)-**91** and (–)-**91** respectively and used as a starting material of the enantioselective synthesis of (–)-boschnialactone **92** [87] (Scheme 47).

A full paper containing all these data (previously published as short communications), as well as some mechanistic proposals concerning the regioselective hydrolysis of the diesters and the unexpected formation under certain conditions of the products not related to the cycloadditions, was published later [88].

The use of dimethyl $(R)_S$-2-(10-isobornylsulfinyl)maleate (**87**) as a chiral synthetic equivalent of dimethyl acetylenedicarboxylate had several limitations arising from (i) the non-trivial preparation of the chiral auxiliary (10-mercaptoisoborneol) required to produce the dienophile, (ii) the low stereoselectivity observed in the synthesis of the thioether used as precursor of the sulfinyl reagent, and (iii) the lack of differentiation of the two ester groups present in the molecule. In order to avoid symmetrization, selective monodemethylation and further re-esterification were required as previous steps of desulfinylation (see

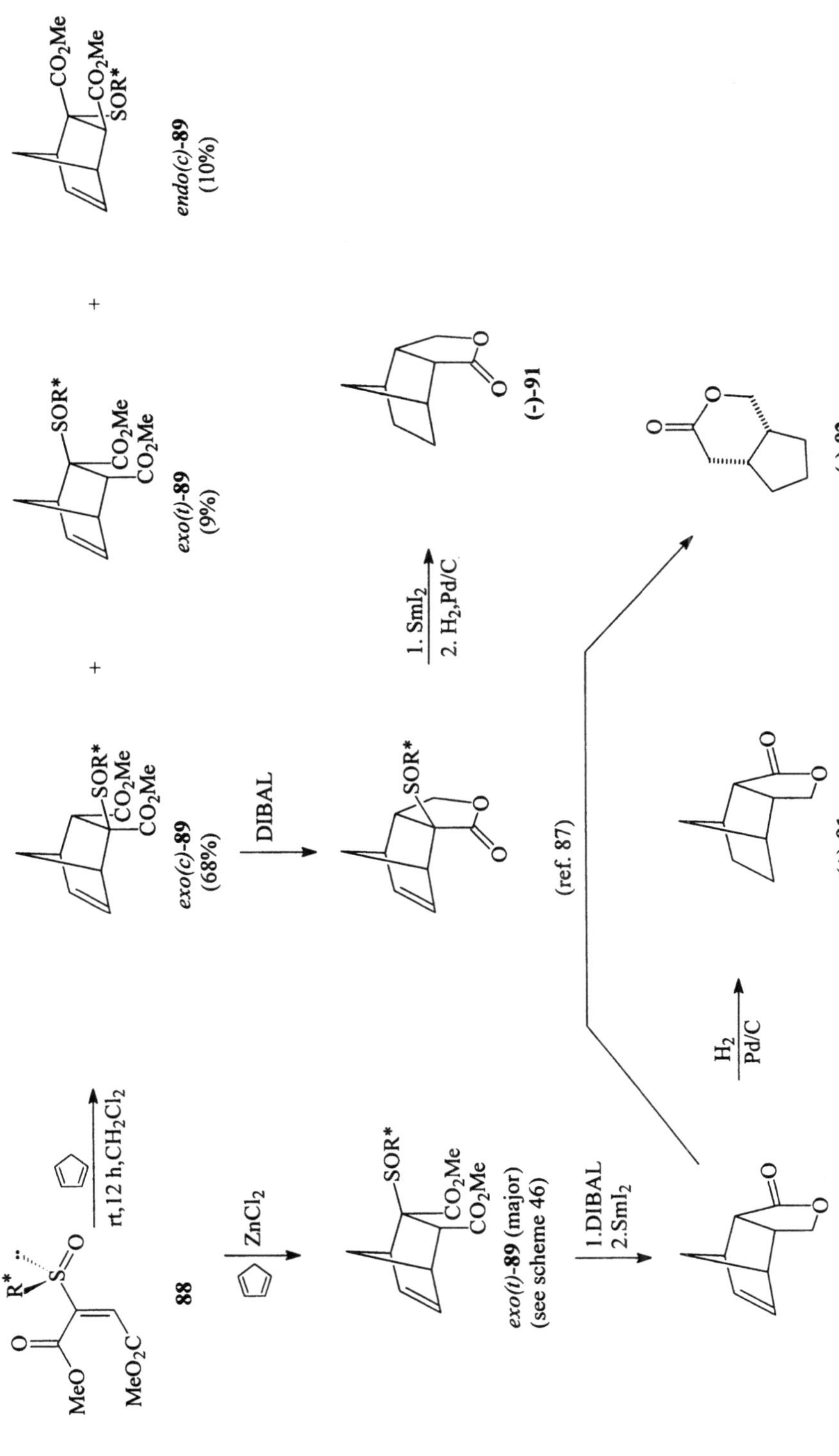
88
rt, 12 h, CH2Cl2
ZnCl2
exo(c)-89 (68%)
exo(t)-89 (9%)
endo(c)-89 (10%)
exo(t)-89 (major) (see scheme 46)
DIBAL
1. SmI2 2. H2, Pd/C
1.DIBAL 2.SmI2
(ref. 87)
H2 Pd/C
(-)-91
(+)-91
(-)-92

Scheme 47

transformation *exo*(*t*)-**89** → in Scheme 46). The synthesis of *p*-tolylsulfinyl maleates with differentiated carboxylic functions was accomplished to overcome these limitations, and so to improve the scope of the sulfinyl maleates as chiral dienophiles [89, 90]. In these papers, the synthesis of the half-esters **93** and **94** as well as the mixed diester **95**, and their reactions with cyclopentadiene under different conditions, were reported. The most significant results from these reactions are collected in Scheme 48. As can be seen, all the crude product mix-

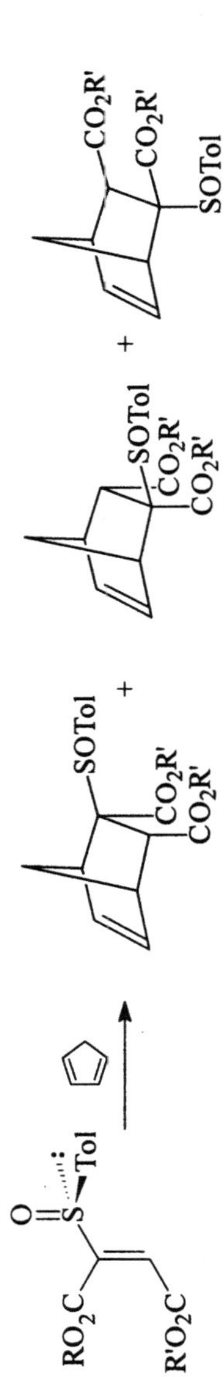

93 R=Bu^t, R'=H
94 R=H, R'=Me
95 R=Bu^t, R'=Me

exo(*c*)-**96** *exo*(*t*)-**96** *endo*(*c*)-**96**
exo(*c*)-**97** *exo*(*t*)-**97** *endo*(*c*)-**97**
exo(*c*)-**98** *exo*(*t*)-**98** *endo*(*c*)-**98**

Dienophile	Catalyst	T (°C)	t (h)	Products *exo*(*c*)	*exo*(*t*)	*endo*(*c*)	Yield (%)
93	-	-20	12	92	5	3	-
94	-	-20	24	70	11	19	-
95	-	rt	41	58	17	25	93
95	$BF_3 \cdot OEt_2$	-20	7	37	43	20	81
95	$ZnBr_2$	-20	7	6	89	5	95
95	Eu(fod)$_3$	-20	12	60	4	36	81

Scheme 48

tures contain the two possible *exo*(*c*) and *exo*(*t*) adducts (corresponding to the *exo*-approach of the diene to the less hindered face of the dienophile in *s-cis* and *s-trans* conformations respectively), but only the *endo*(*c*) adduct (derived from the sterically favored *endo*-approach of diene to the dienophile in its *s-cis* conformation).

All the reactions are *exo*-selective with respect to the sulfinyl group (*exo*/*endo* > 1). In the absence of catalysts they show a π-facial selectivity favoring the formation of *exo*(*c*) adducts. Monoester **93** exhibits the highest *exo*/*endo* and π-facial selectivities. Additionally reactivities of monoesters **93** and **94** are clearly higher than that of diester **95**, which required longer reaction times and higher temperature. Despite these facts suggesting compound **93** to be the most efficient dienophile, the low stability of adducts **96** and **97** derived from monoesters (they decomposed during chromatographic purification) determined that diester **95** was the most suitable to be used in asymmetric Diels-Alder reactions. The most efficient catalysts to improve its reactivity were $ZnBr_2$ and Eu(fod)$_3$. With Eu(fod)$_3$, the π-facial selectivity also increased, but poorer ratios of *endo*/*exo* adducts were obtained. In contrast, with $ZnBr_2$ the *exo*/*endo* selectivity was also improved, whereas the π-facial selectivity, although it is very high, is the opposite to that observed in the absence of catalyst (the *exo*(*t*)-**98** adduct is obtained as the major product).

The stereochemical results observed were explained by assuming that the favored attack of the diene took place at the less hindered face of the dienophile, which adopts the most stable conformation around the C-S bond. This is strongly dependent on the reaction conditions and the substrate structure. In the absence of catalysts, both mono- and diesters mainly adopt the conformation with the sulfinyl oxygen in an *s-cis* arrangement (A in Fig. 7) in order to

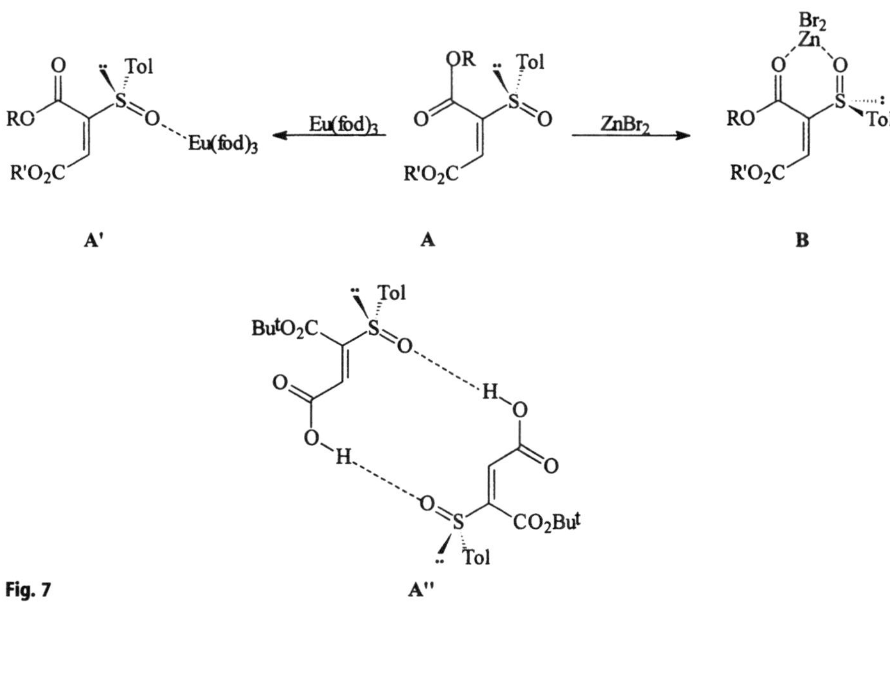

Fig. 7

minimize its electrostatic repulsion with the oxygens of the geminal CO_2R group. Therefore, the major adduct exhibits the *exo(c)* configuration. These *s-cis* conformations can be additionally stabilized by association of the sulfinyl oxygen with $Eu(fod)_3$ (which increases its steric size, see **A′** in Fig. 7), or by intermolecular hydrogen bond formation in the case of the monoester **93** (which would form dimeric species **A″**, and so explain the higher reactivity of its cycloadditions). $ZnBr_2$, acting as a chelating agent, shifts the conformational equilibrium toward *s-trans* conformations (**B** in Fig. 7). The influence of the catalysts on the *endo/exo* selectivity does not appear to be clear.

Problems related to the stability of the $CO_2{}^tBu$ group in reactions of diester **95** conducted under Lewis acid catalysis prompted us to change the alkyl ester group. Diester **99**, containing a benzyl group instead of *t*-Bu, proved a more suitable dienophile, and its reactions with cyclic and acyclic dienes under different conditions were reported [91, 92]. The most significant finding of this new study on cyclic dienes was the fact that $TiCl_4$ is a much more efficient catalyst than others previously used, allowing complete transformation of cyclopentadiene at −78 °C (or −20 °C with cyclohexadiene) in the shortest reaction time. These reactions occur with high levels of *endo/exo* and π-facial selectivity, with the *exo(c)* adducts being the major products. The influence of catalysts such us $Eu(fod)_3$ (affording the highest π-facial selectivity but the lowest *endo/exo* selectivity) and $ZnBr_2$ (inverting the π-facial selectivity in the reaction with cyclopentadiene and thus yielding the *exo(t)* adduct as the major one) remained identical to that observed for compound **95**.

More interesting were the results obtained in reactions of **99** with acyclic dienes catalyzed by $Eu(fod)_3$ (−20 °C) and $TiCl_4$ (−78 °C). The resulting adducts **100** are unstable and underwent spontaneous sulfinyl elimination at room temperature, affording cyclohexadienes. Reactions with dienes lacking substituents at C-1 (butadiene, 2-methyl butadiene and 2,3-dimethyl butadiene) yielded optically pure compounds **101** (Scheme 49). These results indicate that the regioselectivity and the π-facial selectivity of the cycloadditions are complete (only one adduct is formed) under both catalytic conditions. Desulfinylation of **100** is also completely regioselective.

Reactions of **99** with 1-substituted dienes gave a mixture of two different cyclohexadienes **103** and **104**, resulting from the non-regioselective sulfinyl elimination from the unstable adducts **102**. The optical purities of the conjugated cyclohexadienes **103** (*ee* > 96%) are very high regardless of the catalyst used,

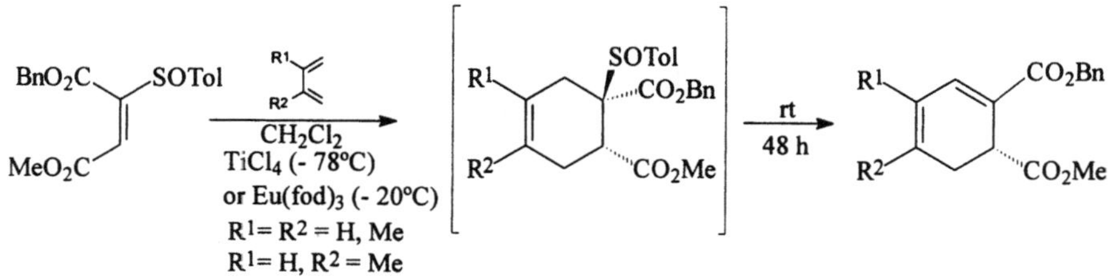

99 **100** **101** (*ee* >96%)

Scheme 49

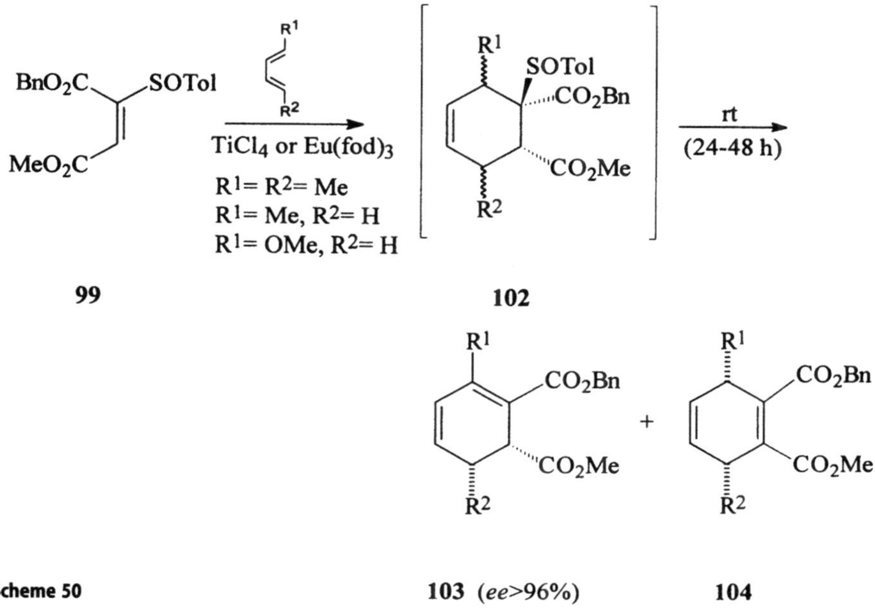

Scheme 50

whereas those of the 1,4-cyclohexadienes **104** are also very high (*ee* > 96%) when they are obtained under $TiCl_4$ catalysis but only moderate (*ee* < 40%) for compounds obtained in the presence of $Eu(fod)_3$ (Scheme 50).

These results were explained by assuming the exclusive formation of the *exo*-**102** adduct in reactions conducted under $TiCl_4$ catalysis, which can undergo transformation into optically pure cyclohexadienes **103** and **104** (Scheme 51). The *endo/exo* selectivity of reactions catalyzed by $Eu(fod)_3$ must be lower (as shown in reactions with cyclic dienes, see Scheme 48), and results in the formation of two adducts, *endo*-**102** and *exo*-**102**. The former can undergo transfor-

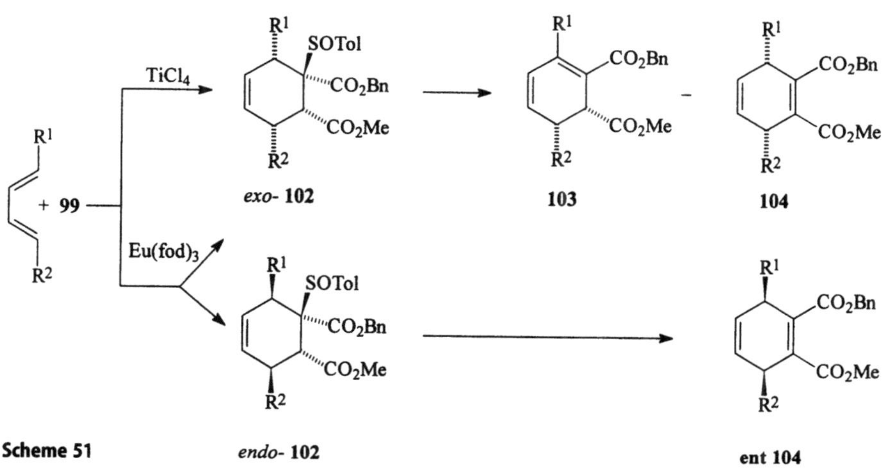

Scheme 51

mation only into conjugated cyclohexadiene (due to the *syn* character of the sulfinyl elimination) yielding ent-**104**, enantiomer of compound **104** (Scheme 51). Therefore, the use of $Eu(fod)_3$ determines a decrease in the *ee* of the conjugated 1,4-cyclohexadiene **104**, but does not affect the optical purity of the 1,3-cyclohexadiene **103**.

The sense of the π-facial selectivity observed in reactions carried out in the presence of $TiCl_4$ was initially explained by assuming the formation of the chelated species **C** (Fig. 8), with the titanium joined to the oxygens of the sulfinyl and ester groups [92]. The orientation and size of the halogens arranged in apical positions of the tetragonal bipyramid centered at titanium make easier the approach of the diene to the bottom face. However further studies made on benzyl 2-*p*-tolylsulfinyl acrylate [46] suggested that both ester groups (rather than one ester group and the sulfinyl group) could be involved in the chelation. Species such as **D** (Fig. 8), with the sulfinyl oxygen in an *s-cis* conformation, would be obtained, thus explaining the formation of the *exo*(*c*) and *endo*(*c*) adducts.

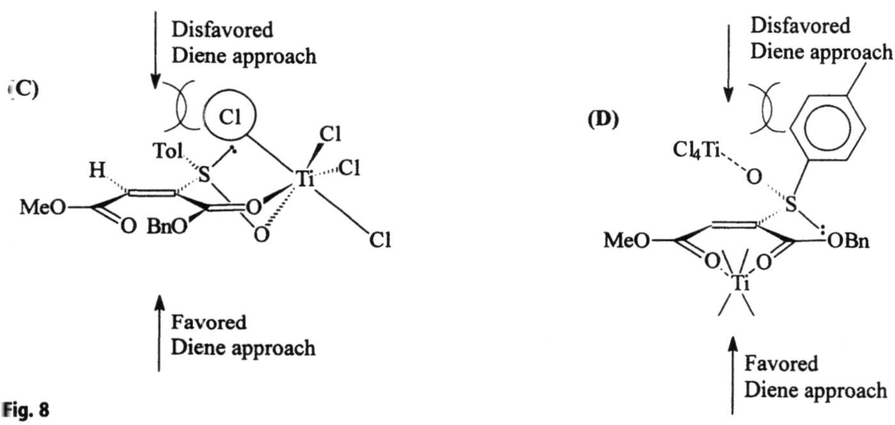

Fig. 8

Based upon the almost complete regio, *endo*, and π-facial selectivities of the reactions of **99** with acyclic dienes in the presence of $TiCl_4$ (only *exo*(*c*)-**102** adducts were obtained), its reactions with 1-vinylcycloalkenes were studied. This study was made to achieve a highly enantioselective approach to functionalized [4.n.0] bicyclic compounds **106** [93]. The adducts **105** resulting in these reactions are unstable and produce mixtures of the two possible sulfinyl elimination products **106a** and **106b** in high optical purity (Scheme 52). The regioselectivity of the elimination depends upon the size of the second ring, the formation of the 1,4-cyclohexadiene derivatives **106b** for seven-membered or larger rings being almost exclusive (Scheme 52). The stereoselectivity of the epoxidation and hydroboration of compounds **106b** was also studied.

Two main problems restricted the synthetic usefulness of the sulfinyl maleates, the low regioselectivity of the elimination of the sulfinyl group in reactions with 1-substituted dienes, and its moderate reactivity (almost identical to that of the sulfinyl acrylates, despite the additional ester group). The use of $TiCl_4$ overcame the second problem, but this catalyst is not compatible with alkoxy substi-

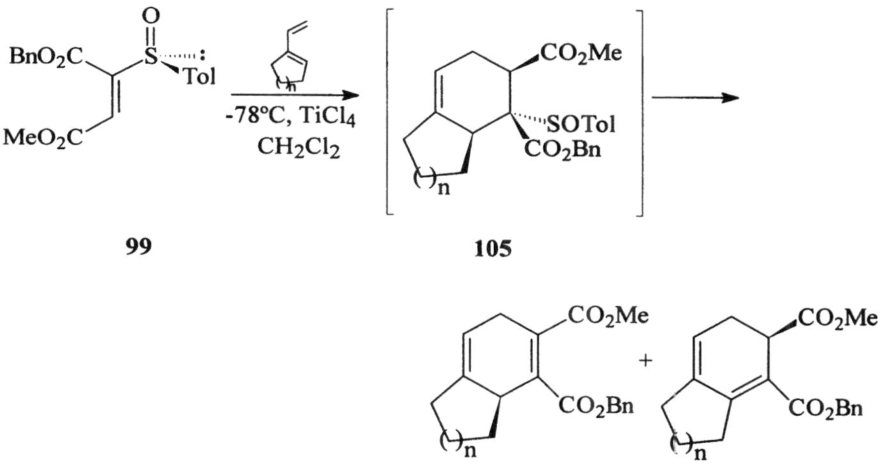

99 105

106a 106b

106a		106b
81	(n=1)	19
61	(n=2)	39
<5	(n=3)	>95
<5	(n=8)	>95

Scheme 52

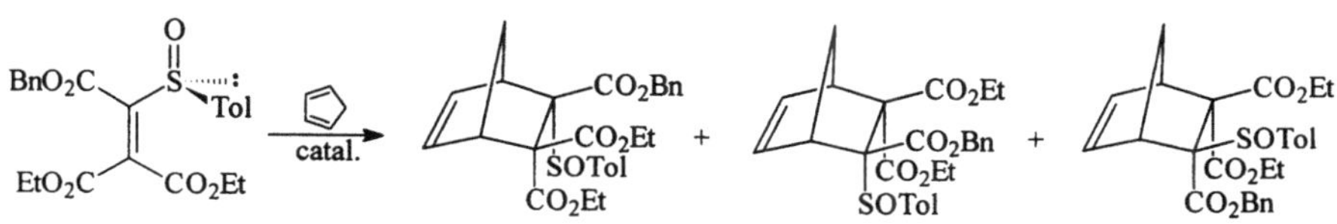

107 *endo(t)*-**108** *endo(c)*-**108** *exo*(t)-**108**

Conditions	catal.	*endo(t)*-108	*endo(c)*-108	*exo*(t)-108
20°C, 72h	-	62	8	30
0°C, 2h	Eu(fod)$_3$	62	8	30
-10°C, 2h	$ZnBr_2$	55	-	45
-78°C, 1h	$TiCl_4$	90	4	6

Scheme 53

tuted dienes (they readily decomposed). The behavior of the sulfinyl trialkoxycarbonyl ethene **107** was investigated in order to determine if the presence of a third ester group was sufficient to surmount both problems [94]. The results obtained indicated that the reactivity of **107** was similar to that of **99**, whereas both π-facial and *endo/exo* selectivities are opposite for each dienophile (Scheme 53). Thus, the *endo*(t) adduct was the major product obtained in reactions from **107** (Scheme 53), whereas it was not detected in reactions starting from **99** (Scheme 48). In contrast, the *exo*(c) adduct was predominant in the reaction mixture obtained from **99**, but it was not formed from **107**. The reactions of **107** with acyclic dienes yielded 1,3-cyclohexadienes with low optical purity (despite the presumably almost complete π-facial selectivity of these cycloadditions) which was attributed to a decrease in the *endo/exo* selectivity. These reactions were also investigated under high pressures. The results obtained were almost identical indicating the influence of the pressure in the *endo*-selectivity is scarcely significant and therefore the optical purity of the adducts remained very low.

Maleimides have high dienophilic reactivity, and the synthesis of interesting chiral sulfinyl maleimides and their use as powerful dienophiles for asymmetric Diels-Alder reactions have also been reported. Thus, enantiomerically pure *N*-alkylsubstituted α-(2-*exo*-hydroxy-10-bornylsulfinyl) maleimides (**109**) underwent Diels-Alder reactions catalyzed by $ZnCl_2$ with cyclopentadiene and furan affording the corresponding cycloadducts with high diastereoselectivities [95]. In this communication the authors claim these sulfoxides as the first practically useful chiral sulfinyl dienophiles (they react diastereoselectively with poor Diels-Alder dienes, such as furan, in relatively short periods of time). The reactions of **109** with cyclopentadiene under $ZnCl_2$ catalysis proceeded with high diastereofacial selectivity (*de* ≈ 94% with the *exo*-(*t*) adduct favored, when R = Me, CH_2Ph) and complete *endo/exo* selectivity in all cases (*exo* sulfinyl adducts are the only one detected). Further studies on other maleimides [96] confirmed the almost complete π-facial (*de* ranged between 88 and 99% for different R groups) and *endo/exo* selectivities of these reactions, as well as the inversion of the former in the absence of the $ZnCl_2$ (*de* ≈ 45%, *exo*(*c*) being favored). The reactions with cyclohexadiene are even more stereoselective (Scheme 54).

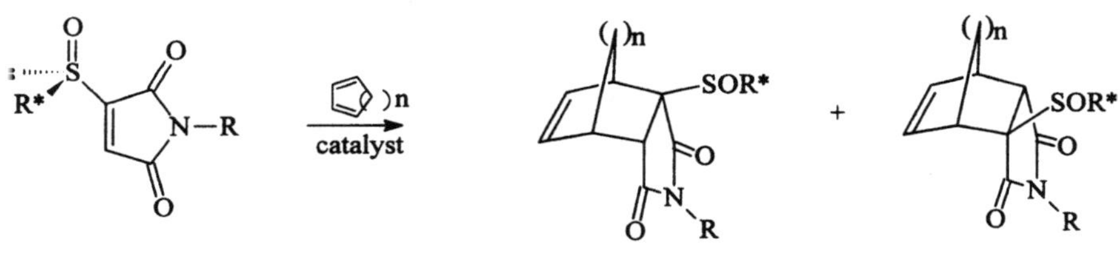

109 *exo(t)* *exo(c)*

(R = Me,Ph,CH_2Ph)

R* = 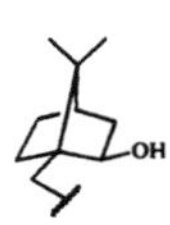

Catalyst	n	*exo(t)*	*exo(c)*
–	1	27-30	73-70
–	2	22	78
$ZnCl_2$	1	94-99.5	6-0.5
$ZnCl_2$	2	100	0

Scheme 54

The reactions of **109** with furan catalyzed by $ZnCl_2$ gave different results depending on the reaction temperatures [95, 96] (Scheme 55). At 0 °C the π-facial selectivity for both *endo* and *exo* addition modes is complete (only *endo*(*t*) and *exo*(*t*) adducts are obtained) but the *exo/endo* ratio is very low (*de* = 36%), *endo*(*t*) being the major adduct. At room temperature, the reactions exclusively afforded *endo*(*c*) and *endo*(*t*) adducts by dissociation (retro-Diels-Alder) and

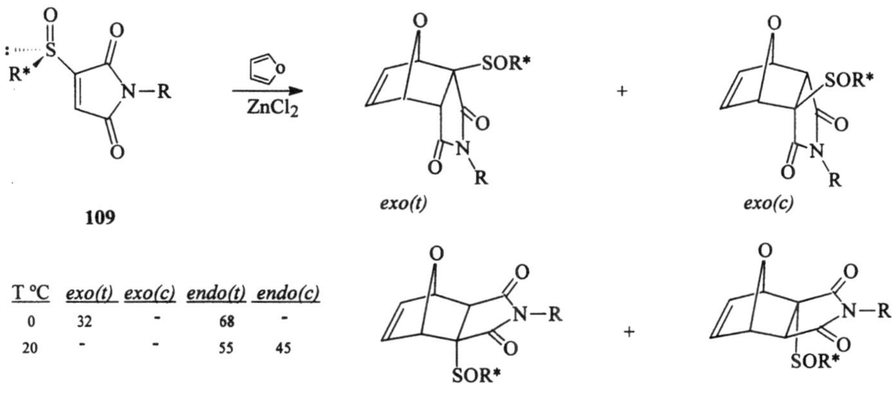

T °C	exo(t)	exo(c)	endo(t)	endo(c)
0	32	-	68	-
20	-	-	55	45

Scheme 55

recombination, with exclusive formation of the thermodynamically most stable *endo* adducts (which are *exo* with respect to the maleimide moiety). In the absence of $ZnCl_2$, almost identical amounts of the four possible adducts were obtained, thus demonstrating that the π-facial selectivity is very low under these conditions (but not inverted as observed in reactions with cyclopentadiene).

The high diastereofacial selectivity observed in catalyzed reactions was explained by assuming the formation of the chelated species **A** (Fig. 9), where the metal is coordinated to the sulfinyl and imido carbonyl oxygens (the sulfinyl oxygen adopts an *s-trans* conformation). This species will be preferentially attacked by dienes from the less hindered, lone-pair, face (Fig. 9).

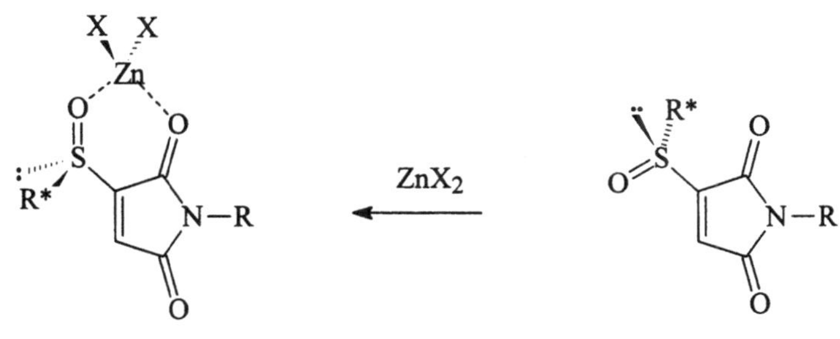

Fig. 9 **A** **B**

Nevertheless, the authors do not comment on the different *endo/exo* selectivity observed in reactions with furan (the *endo* adduct is predominant, Scheme 55) and cyclopentadiene (only *exo* adducts are obtained, Scheme 54) under kinetic control conditions. They explain the changes in the facial diastereoselectivity, observed in the absence of $ZnCl_2$, as a consequence of the increase in population of conformation **B** (Fig. 9), attributed to dipole-dipole repulsion. However they pay no attention to the role of the diene, which could be mainly responsible for the observed differences. With furan, the *de* decreased, but the major adduct remains the same, whereas with cyclopentadiene, the π-facial selectivity is clearly inverted [46]. All these facts suggest a special role for furan not discussed so far in the literature.

By using a sequence involving regio and diastereoselective reduction of the adducts (it only affects the carbonyl far away from the sulfinyl group, yielding γ-oxygenated lactones), stereoselective *N*-acyliminium addition, and retro Diels-Alder reactions as the main key steps, Koizumi et al. have synthesized chirally functionalized pyrrolidines [96, 97], pyrrolizidines [98], and indolizidines [98–100], depicted in Scheme 56. The milder conditions required for the evolution of the adducts derived from furan preserve the chirality of the substrates.

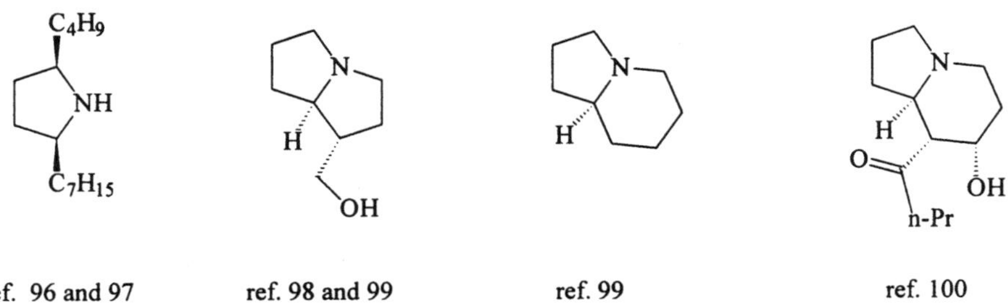

Scheme 56

2.1.3.2
Sulfinyl Quinones

The first papers related to the use of sulfinyl quinones as dienophiles concerned racemic naphthoquinone derivatives, obtained by MCPBA oxidation of their corresponding sulfides. They were focused to clarify the role of the sulfinyl group in the regioselectivity of the Diels-Alder reactions as well as to take advantage of its ready pyrolytic elimination for synthetic purposes. Hence, in 1978 Boeckman et al. [101] explored the reactions of relatively unpolarized dienes with different juglone derivatives as dienophiles (including thioethers, sulfoxides **110** and **111**, and sulfones) in order to prove that the electronic structure of the diene component was much more important for the regiocontrol than the dienophile structure. The results obtained from sulfoxides (Scheme 57) evidenced a significant influence of the sulfinyl group on the regioselectivity (able to overcome that of the hydrogen bond between the OH and the carbonyl quinonic group). These results were applied later to a series of model studies leading toward the synthesis of the anthracyclinones antibiotics adryamicin and daunorubicin [102].

The ready evolution of the adducts into aromatic quinones by spontaneous sulfinyl elimination and further aromatization prompted the use of sulfinyl naphthoquinones as a synthetic equivalent of the unknown compound naphthynoquinone [103]. For this purpose, sulfinyl quinones represent a convenient synthetic alternative to haloquinones. The highly regioselective course of the Diels-Alder reactions of 2-phenylsulfinyl-1,4-naphthoquinones (as well as their corresponding thioethers and sulfones) unsymmetrically substituted by

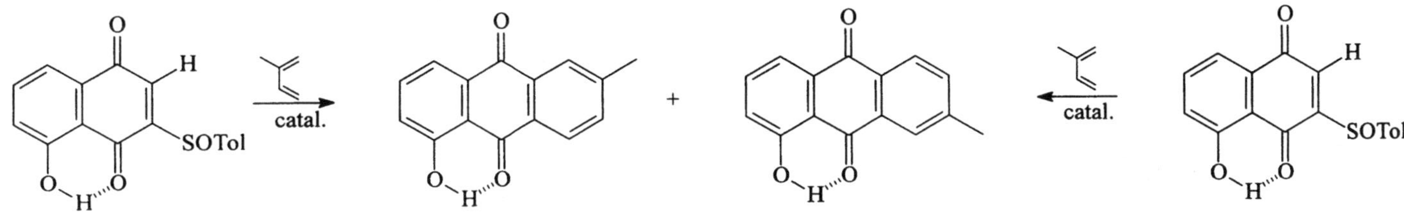

110

111

Dienophile	Catal.		
110	-	2.9	1
	BF_3OEt_2	5.4	1
111	-	1	2
	BF_3OEt_2	1	4.3

Scheme 57

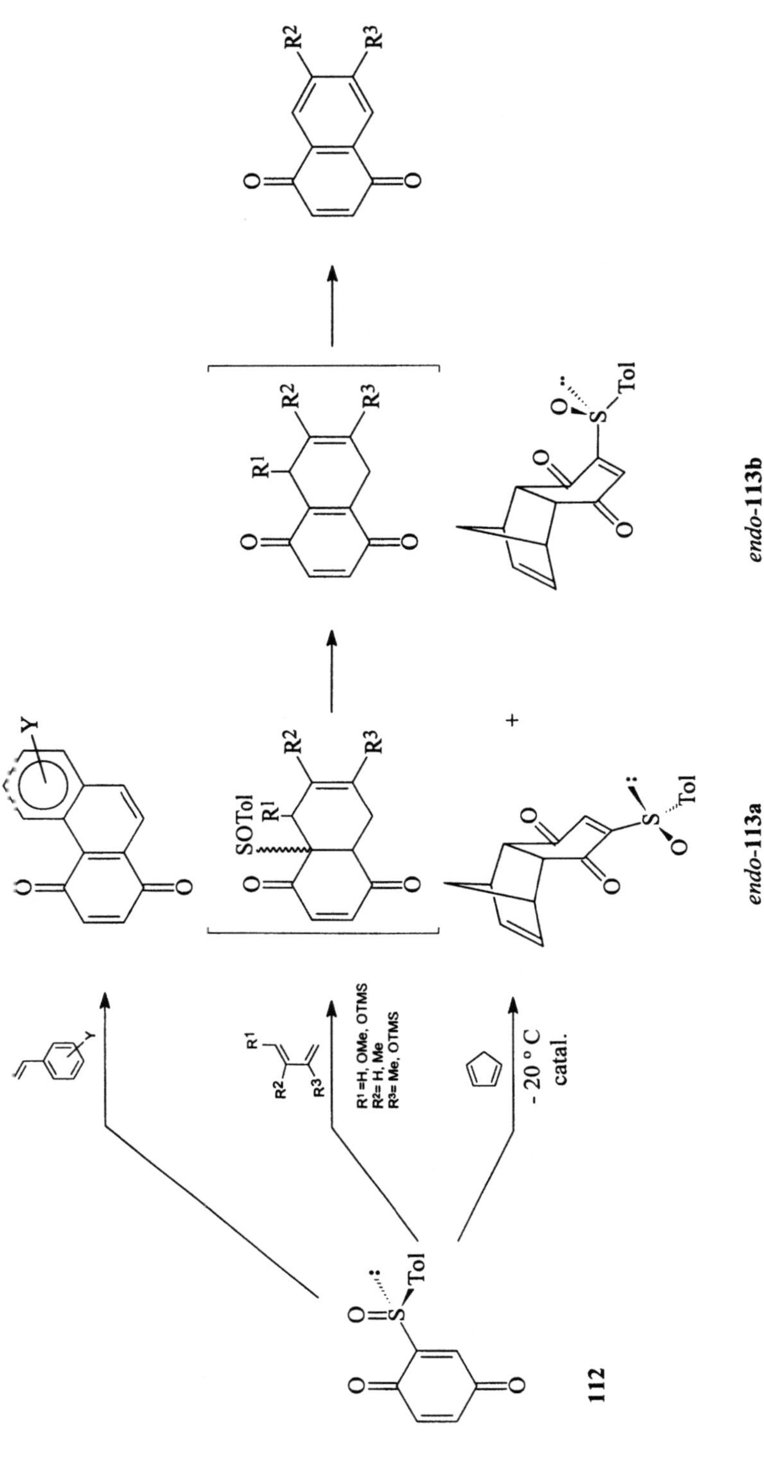

Catal.	*endo*-113a	*endo*-113b
-	72	28
Eu(fod)$_3$	89	11
BF_3OEt_2	10	90

Scheme 58

methoxy groups on the benzenoid ring, was utilized to synthesize several anthraquinones, including 11-deoxyanthracyclinones and natural products such us pachybasin and phomarin 6-methyl ether [104].

The first synthesis of enantiomerically pure sulfinylquinone and its use in asymmetric Diels-Alder reactions were reported in 1989 [105]. (*S*)-2-*p*-Tolylsulfinyl-1,4-benzoquinone (**112**) reacted with cyclopentadiene (Scheme 58) yielding a mixture of the adducts (*endo*-**113a** and *endo*-**113b**) resulting in the *endo*-approach of diene to the two diastereotopic faces of the unsubstituted C_5-C_6 double bond at the starting quinone. A complete *endo*-selectivity and a significant π-facial selectivity were observed. The use of Eu(fod)$_3$ as a catalyst increased the π-facial selectivity, whereas the addition of BF_3 caused an inversion in the sense of the facial selectivity (*endo*-**113b** is obtained as the major adduct, Scheme 58). With acyclic dienes, the reaction takes place on the sulfinyl substituted double bond C_2-C_3, but it was not possible to obtain any stereochemical information due to desulfinylation and further aromatization of the initially formed adducts (Scheme 58). Nevertheless, this development has been used to synthesize a wide range of substituted 1,4-phenanthrenequinones by reaction of **112** with substituted styrenes [106] (Scheme 58).

Different strategies have been used to synthesize optically pure 2-*p*-tolylsulfinyl benzoquinone **112**. The best involves sulfinylation of the hydroquinone dimethylether or its brominated derivative, followed by CAN oxidation to give the sulfinyl quinone (Scheme 59) [107a]. Ortholithiation or metal-halogen exchange of the starting materials were respectively used as the source of the arylcarbanions acting as nucleophiles in reactions with menthyl sulfinate.

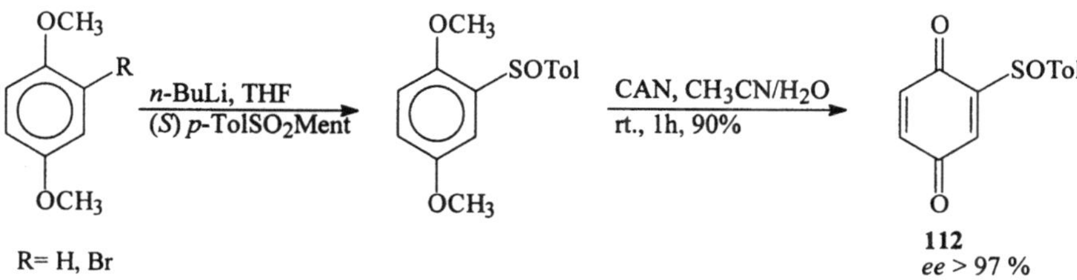

Scheme 59

A similar route was used to prepare 2-*p*-tolylsulfinyl naphthoquinone **114a** [107b], its derivatives **114b**, **114c**, and **120** [107b, 108] containing methoxy and hydroxy substituents in different positions (the ring with higher electronic density is the most easily oxidized by CAN), as well as 3-ethyl (**124**) and 3-chloro (**125**) 2-*p*-tolylsulfinyl benzoquinones [109].

The reaction of sulfinyl naphthoquinones **114a–c** with cyclopentadiene afforded mixtures of two *exo*-sulfinyl adducts (the *endo*-orientating character of the quinonic system is clearly predominant). Compound *exo*(*c*)-**115** was the major product in CH_2Cl_2 at −20°C (*de* ranged between 80 and 90% depending on the dienophile). The π-facial selectivity of the process was reversed in the presence of $ZnBr_2$, *exo*(*t*)-**115** becoming predominant or exclusive [110]. These catalyzed cycloadditions required shorter reaction times and took place with very high π-facial selectivity (*de* > 97%) with dienophiles **114a** and **114b**, but much lower (*de* ~ 20%) with **114c** (Scheme 60). Compounds *exo*(*c*)-**115** and *exo*(*t*)-**115** are thermally unstable and undergo transformation at room temperature into desulfinylated compounds **117** and **117′**, respectively (identical starting from **114a**, but enantiomeric from **114b** and **114c**).

Similar results were obtained in reactions of **114a–c** with cyclohexadiene. As a consequence of the lower reactivity of this diene, boiling $CHCl_3$ was required to achieve complete transformation of the dienophile in the absence of catalysts, which precluded the isolation of the initial adducts *exo*(*c*)-**116** and *exo*(*t*)-**116**, and led directly to desulfinylated compounds **118** and **118′** (identical or enantiomers, see above). In $ZnBr_2$ catalyzed cycloadditions, the required reaction conditions are milder, which allows detection of adducts *exo*(*c*)-**116** and *exo*(*t*)-**116** by NMR spectroscopy. This revealed that their π-facial selectivity was higher than that obtained from cyclopentadiene.

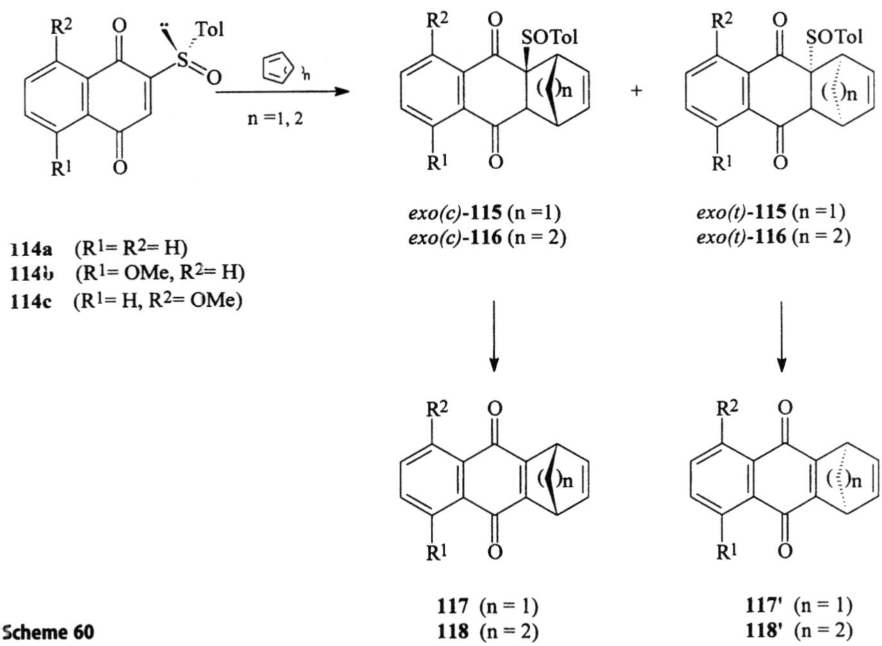

Scheme 60

The study of the tautomeric equilibrium of sulfinyl naphthazarin **119** indicated an almost identical participation of both forms, **119A** and **119B** [108]. The reaction of this compound with cyclopentadiene at –30 °C yielded a 77 : 23 mixture of adducts **120** (the two diastereoisomers resulting from the evolution of **119A**) and the quinone **121** (resulting from desulfinylation of the adduct obtained by reaction of **119B**) (Scheme 61). Only the adducts **120** were obtained in reactions catalyzed by BF_3, explained as the result of an increase in the relative reactivity of **119A** in the presence of the catalyst. Some contribution of a shifting of the tautomeric equilibrium toward **119A**, promoted by the association of the catalyst with the sulfinyl oxygen (Scheme 61), is also suggested [111].

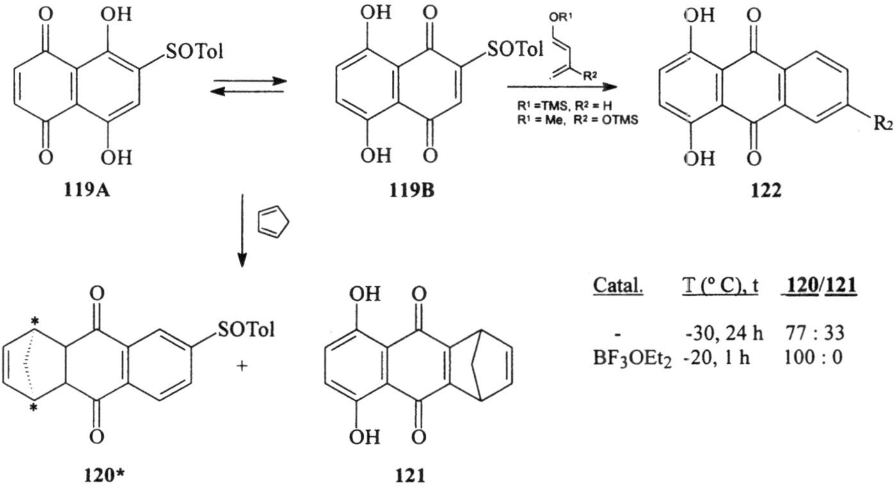

Catal.	T (° C), t	120/121
-	-30, 24 h	77 : 33
BF_3OEt_2	-20, 1 h	100 : 0

* (*de* ≅12-40%)

Scheme 61

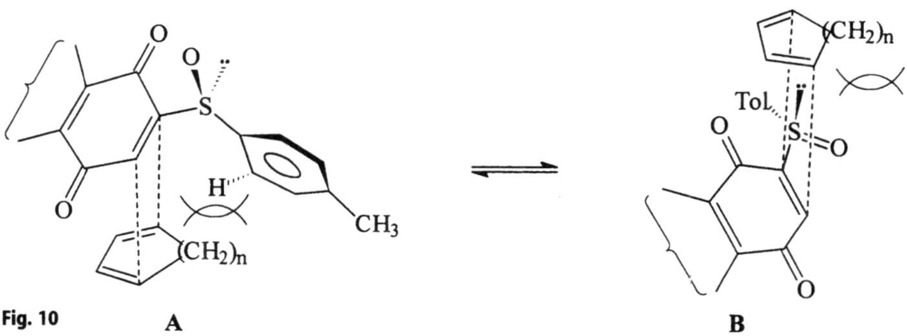

Fig. 10

These results suggested that the reactivity of the unsubstituted dienophilic double bond in **119A** is higher than that of the sulfinylated double bond in **119B**. A similar situation had been found in the reaction of 2-*p*-tolylsulfinyl *p*-benzoquinone **112**, which reacted unexpectedly with cyclopentadiene at C_5-C_6 instead of the sulfinylated C_2-C_3 double bond (Scheme 58). The negative influence of the sulfinyl group of the dienophilic reactivity toward cyclopentadiene was not expected upon consideration of the electron-withdrawing character of the sulfinyl group. It was explained by assuming that destabilizing steric interactions between the methylene bridge of cyclic dienes and the substituents around the sulfur during the *exo*-sulfinyl approach must prevail over the beneficial electronic effect of the SOTol group. In Fig. 10 it can be seen that these interactions take place mainly with the substituent in *s-cis* arrangement with respect to the dienophilic double bond, which must determine a different reactivity of A and B conformations. The results obtained with acyclic dienes (lacking methylene bridge) support this explanation, as the sulfinylated double bonds are clearly more reactive than those without substituents. Thus, compound **119** yielded exclusively aromatic derivatives **122**, resulting in desulfinylation and further alkoxy elimination from the adducts formed by attack of the diene on the tautomer **119B** (Scheme 61), whereas 112 afforded only compounds derived from the adducts obtained by reaction of the sulfinylated double bond C_2-C_3 (Scheme 58).

The regioselectivity of the cycloadditions on the sulfinylated double bonds in sulfinyl quinones was also investigated by studying reactions of compounds **114a–c** with 1-methoxy-1,3-cyclohexadiene [110]. Only optically pure compounds **123** were obtained, thus indicating that the sulfinyl group exerts complete control over both regioselectivity and π-facial selectivity (Scheme 62). It is noteworthy that, for compound **114c**, the orientating effect of the sulfinyl group on the regioselectivity clearly prevails over that of the 5-methoxy substituent (both groups polarize the dienophilic double bond in the opposite sense), and hence the formation of only one regioisomer is observed.

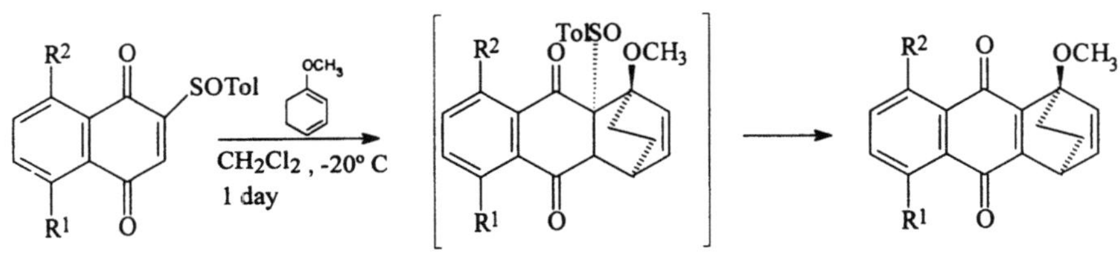

114a (R1= R2= H)
114b (R1= OMe, R2= H)
114c (R1= H, R2= OMe)

123 a-c (*ee* >97%)

Scheme 62

Diels-Alder reactions of 3-chloro and 3-ethyl-2-*p*-tolylsulfinyl-1,4-benzoquinones (**124** and **125**) take place at the C_5-C_6 unsubstituted double bond with both cyclic and acyclic dienes [109]. This is not unexpected from the electronic and steric effects of the substituents at C-3, which makes more difficult the approach of the dienes (even the acyclic ones) to C_2-C_3 than to C_5-C_6. With cyclopentadiene, no π-facial selectivity is observed under thermal conditions, and it is moderate in the presence of $ZnBr_2$ (Scheme 63). Slightly higher π-facial selectivity was observed in reactions with 2,3-dimethyl-1,3-butadiene, presumably due to the lower reactivity of the diene.

One of the most interesting and intriguing findings of the cycloadditions to the unsubstituted double bonds of these quinones is their significant π-facial selectivity. The *de* ranged between 40% and 72% for reactions of **124** and **125** catalyzed by $ZnBr_2$ (Scheme 63), between 44% and 80% for reactions of **112** with cyclopentadiene under different conditions (Scheme 58), and between 12% and 40% for reactions of **119** with cyclopentadiene (Scheme 61). Although a steric approach control was initially postulated to explain the results obtained for **112** [105], the large magnitude of the facial discrimination was surprising considering the long distance existing between the reactive double bond and the chiral

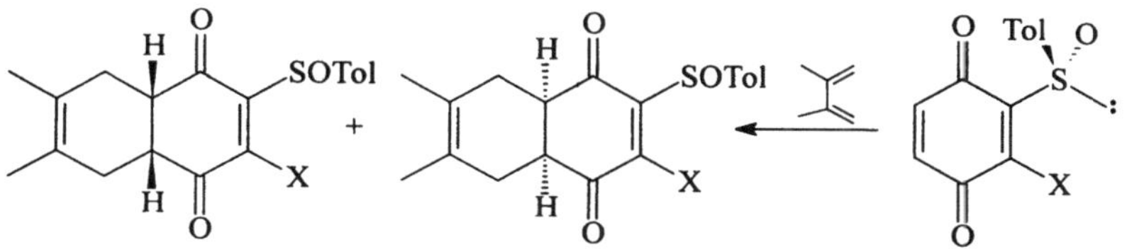

		Catal.	
67	33	-	**124** (X=Cl)
86	14	$ZnBr_2$	
60	40	-	**125** (X=Et)
70	30	$ZnBr_2$	

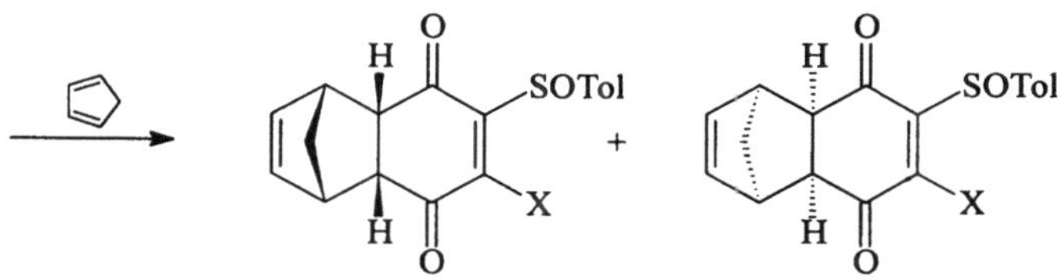

Catal.		
-	50	50
$ZnBr_2$	80	20
-	50	50
$ZnBr_2$	70	30

Scheme 63

sulfinyl inductor. This would exclude or minimize the role of the steric effects in the control of this remote asymmetric induction, thus suggesting it could be a consequence of stereoelectronic factors independent of the distance. Desymmetrization of the π-cloud of the quinonic system, due to its electronic repulsion with the lone electron pair at sulfur, has been recently invoked [112, 113] to explain this behavior. The magnitude of such repulsion must be related to the spatial arrangement of the interacting electrons (i.e., the conformation around the C-S bond) and will be less marked when the π-extended system is larger (as in sulfinylnaphthazarin **120**). According to this explanation, the electronic density of the π-cloud will be higher on the opposite face to that displaying the unshared electron at sulfur. The preferred attack of electron rich diene (acting as nucleophile) should occur from the electron poorest face of the quinone (acting as an electrophile). As depicted in Fig. 11, the favored approach of diene must take place from the bottom or the upper face depending on whether the conformation is A or B.

The stereochemistry of the adduct **113a**, obtained as major under thermal conditions, indicates that it is derived from *endo*-approach to conformation A. It suggests that rotamers with the sulfinyl oxygen in *s-cis* arrangement (B in Fig. 10) are not the most populated in the conformational equilibrium, even though they are favored upon electrostatic grounds. The higher stability of con-

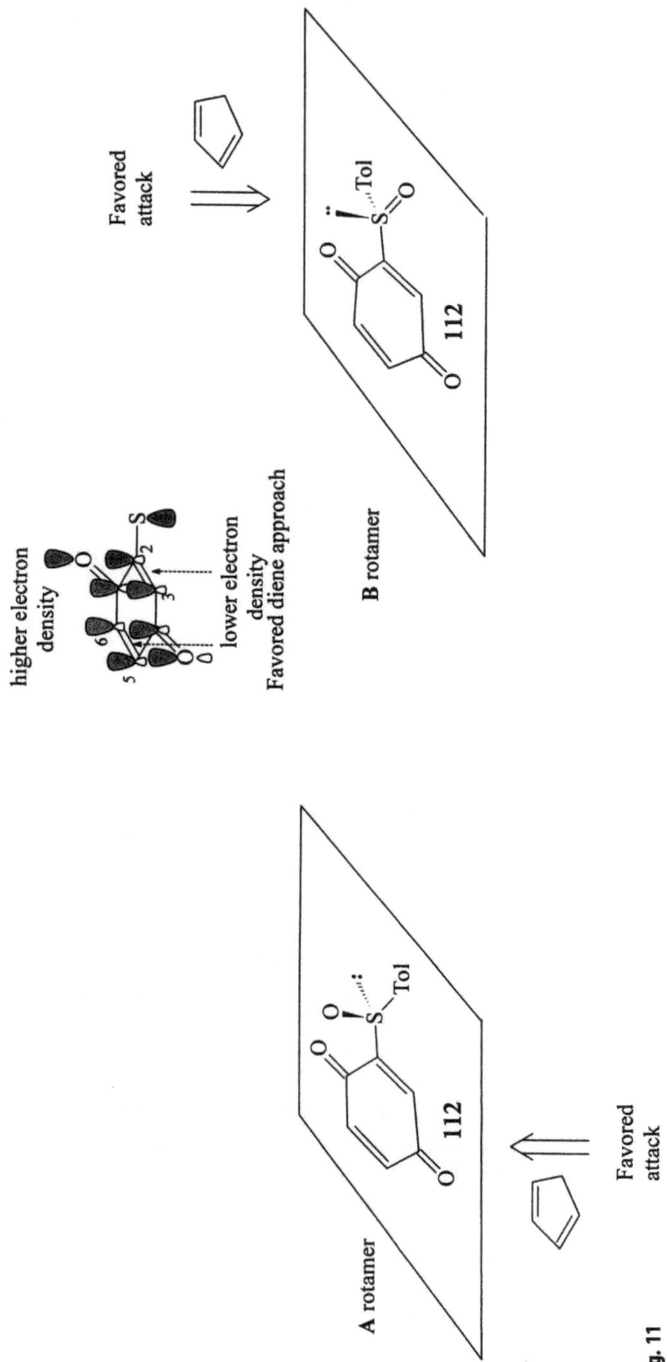

higher electron density
lower electron density
Favored diene approach
Favored attack
112
Tol
B rotamer
A rotamer

Fig. 11

formations such as A (Fig. 10) was explained by assuming a stabilizing $n^2 \rightarrow d^0$ donor-acceptor interaction (not possible for **B** conformations) between the lone electron pair at quinonic oxygen and the empty *d* orbital at sulfur oriented toward the carbonyl [110, 113]. The influence of the reaction conditions (mainly polarity of the solvents and catalysts) on the conformational preferences around the C-S bond are consistent with the observed changes in the composition of the reaction mixtures. At this point, we must indicate that the asymmetric induction shown in Fig. 11 is just the opposite to that suggested by Kahn et al. [17, 18] which postulated that the nucleophilic face of dienophile (the richest electron face) would be that supporting the lone electron pair at sulfur. According to this proposal, the π-facial selectivity of these cycloadditions could also be explained by assuming that rotamer **B** is the most populated one. However, this explanation is not consistent with the influence of the solvent on the stereoselectivity.

Concerning the π-facial selectivity of the cycloadditions on the sulfinylated C_2-C_3 double bond, the available results suggest that this is mainly governed by steric interactions (which in this case prevail over the previously described electronic ones), conferring the highest reactivity upon conformation **B** (Fig. 10). Interactions of dienes, especially the cyclic ones, with the *p*-tol group in conformations **A** must be strongly destabilizing (see Fig. 10). Therefore, these reactions yield the *exo*(*c*) adducts as the major products (Fig. 12), even in the case that *s-cis* **B** were not the most populated conformation (see above). This situation may become different in reactions catalyzed by $ZnBr_2$. In the presence of this catalyst, able to form weakly chelated species **A**, the favored approach of cyclopentadiene or cyclohexadiene takes place toward the less hindered bottom face of the chelated *s-trans* **A′** rotamers (Fig. 12), including the formation of the *exo*(*t*) adducts as the major products and therefore inverting the π-facial selectivity with respect to that observed in the absence of catalysts. The reactions with C-1 substituted dienes (1-methoxy-1,3-cyclohexadiene or piperylene) yielded the *exo*(*c*) adducts as the major products (as in the absence of catalyst) suggesting that the reactivity of the associated (but non-chelated) s-*cis* **B′** rotamers is higher than that of the chelated s-*trans* **A′** conformers. Destabilization of the TS yielding *exo*(*t*) adduct due to steric interactions of the substituent **R** with the metal bridge (Fig. 12), has been invoked to explain this behavior.

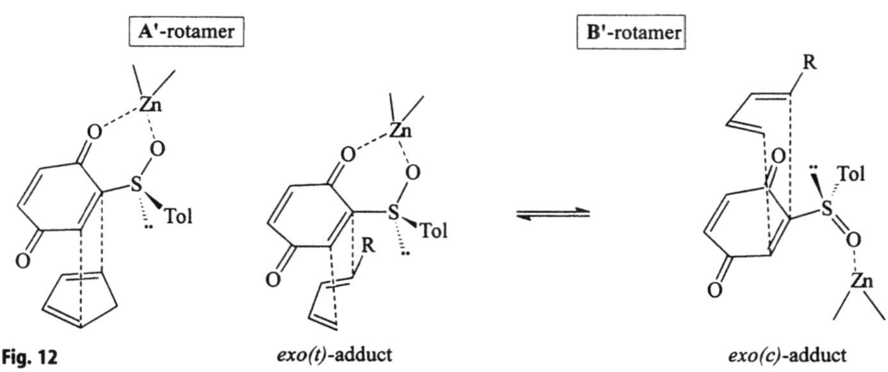

Fig. 12

The results obtained in Diels-Alder reactions of the only sulfinylquinonimine so far reported, *N*-(*tert*-butoxycarbonyl)-3-*p*-tolylsulfinyl-1-benzoquinone-4-imine [114], are completely similar to those obtained from its corresponding sulfinylquinone **112**. Thus, reactions with cyclopentadiene take place at the unsubstituted C_5-C_6 double bond under thermal or Eu(fod)$_3$ catalyzed conditions, or at C_2-C_3 in the presence of $ZnBr_2$. The reactions are completely *exo* selective, and the π-facial selectivities are similar to those reported for sulfinyl benzoquinone **112**. Reactions with piperylene take place at the sulfinylated double bond with complete regioselectivity controlled by the sulfinyl and/or the imine groups acting in a matched way. The resulting adducts are transformed in situ into the desulfinylated compound, whose *ee* has not been reported (this precludes evaluation of the π-facial selectivity of this reaction).

Adducts *endo*-**113a** and *endo*-**113b**, obtained by reaction of sulfinyl benzoquinone **112** with cyclopentadiene (Scheme 58), have proved to be adequate rigid models to evaluate the ability of the sulfinyl group to control the diastereoselectivity of the [4+2] cycloadditions of cyclopentadiene at the ene-dione moiety [115]. The results obtained were explained by taking into account the influence of the association between the sulfinyl oxygen and the different Lewis acids on the conformational equilibrium around the C-S bond. In this sense, the results obtained reinforce the assumptions established in previous papers concerning the role of different catalysts. Moreover, the lack of planarity of the enonic systems used as the starting materials decreases their *endo*-orientating character, which is clearly lower than that of their precursor quinones. It is therefore possible that *exo-anti-endo*-**126** or *endo-anti-endo*-**126** bisadducts as the major products depend upon the experimental conditions (in Scheme 64 only reactions of **113b** are depicted). When a quinonic structure was restored by transformation of the monoadducts **113** into **127**, these new dienophiles reacted with cyclopentadiene in a completely *endo*-selective manner. The π-facial selectivity of these reactions was only dependent upon the sulfinyl group, and it was possible to observe opposite diastereoselection under thermal conditions and in the presence of $ZnBr_2$. In Scheme 64 we have only depicted the reactions connecting **113b** with *endo-anti*-**128b** and *endo-syn*-**128b**. As a similar behavior was observed for **113a**, these reactions allowed the highly stereoselective production of the four possible *endo*-adducts **128**, which are optically pure synthetic equivalents of norborneno-*p*-benzoquinone-cyclopentadiene bisadducts [116] (Scheme 64).

Concerning reactions of acyclic dienes with enantiomerically pure sulfinylquinones, the most interesting contributions have been reported in the last two years. The reaction of **114a** with 2 equiv. of the racemic vinylcyclohexenes (±)-**129**, bearing allylic oxygenated substituents on the cyclohexane ring, afforded only the enantiomerically pure or highly enriched angularly tetracyclic quinones **130A** (>97% ee for $R' \neq H$, see Scheme 65), resulting in a tandem cycloaddition/pyrolytic sulfenic acid elimination [117a,b]. The higher reactivity of the enantiomer (+)-**129** provokes the kinetic resolution of the racemic dienes, thus resulting in the recovery of (–)-**129** (50% ee). Treatment of **130A** (R^2=Me, R^1=TBS) with DBU afforded optically pure anthraquinone **131** [117a]. This is the first asymmetric approach to the tetracyclic skeleton of angucyclinones

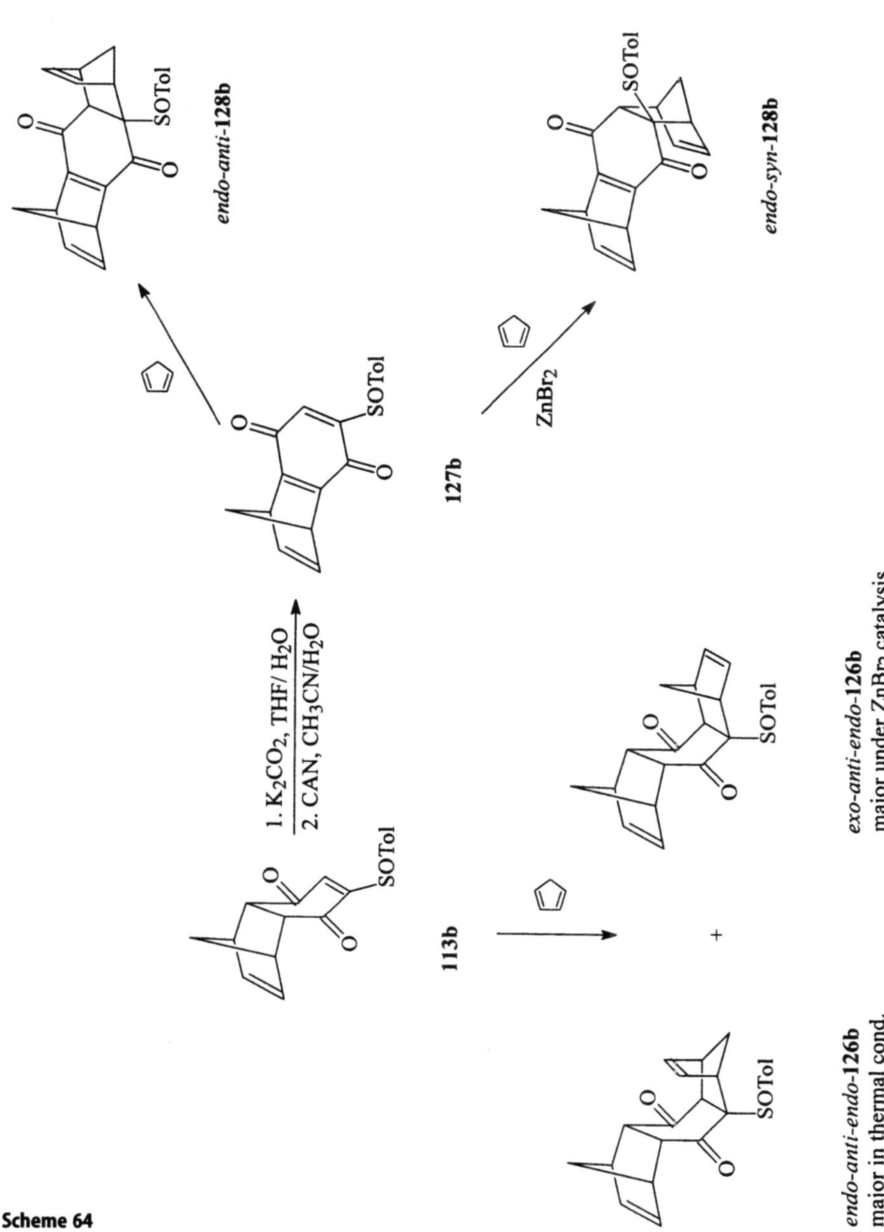

Scheme 64

from chiral dienophiles. Steric effects and torsional interactions account for the observed π-facial selectivities as well as for the configuration of the predominant enantiomer (–)-**129** in the recovered diene.

Similar results were obtained in reactions of **114a** with acyclic dienes bearing oxygenated functions at the allylic stereogenic carbons [117c]. In this case, the reactivity of both enantiomers of diene are not so different and thus 1:3 mixtures of diastereoisomers **130′A** and **130′B**, epimers at the hydroxylic carbon, were obtained. Transition states TS_A and TS_B yielding adducts, whose pyrolytic sulfenic acid elimination respectively afford **130′A** and **130′B**, are depicted in Scheme 65. As we can see, both transition states correspond to the favored approach of each diene's enantiomer toward the upper face of the quinone in *s-cis* arrangement. Moreover, they agree with the rule that the conformers of the diene must be staggered with respect to forming bonds and direct the hydrogen (the lowest sized substituent) toward the sulfur function at dienophile. TS_B is slightly more stable than TS_A on steric grounds (the R group is larger than the OR one, and they interact with quinone residues in both transition states), thus justifying the predominance of **130′B** in the reaction mixture. As the cyclic structure of vinyl cyclohexenes **129** would only allow the formation of transition states such as TS_A (see Scheme 65), the exclusive formation of **130A** is observed, and the kinetic resolution of diene is more efficient.

Starting from 2-sulfinyl benzoquinones, other synthetic applications taking advantage of the tandem cycloaddition/pyrolytic sulfenic acid elimination reactions have also been reported. The reactions of **112** (R = H, Me) with the Dane's diene in the presence of $ZnBr_2$ yielded regioisomers (–)-**132B** (Scheme 66) with enantiomeric excess of 36% (R = H) and >97% (R = Me). By contrast, in the absence of catalyst, compound (+)-**132A** (80 % ee) is exclusively obtained, thus providing an easy and regiocontrolled entry to optically active tetracyclic quinones [118a]. The origin of the reversed regiochemistry is not clear, but it seems to be associated with changes in the coefficients of the diene HOMOs due to the coordination of the catalyst with the methoxy group.

Reactions of **112** (R = H) with divinyl benzenes and naphthalenes have allowed the enantioselective synthesis of helicenebisquinones in high optical purity (30–88% ee) [118b]. In Scheme 66 is also depicted the results obtained from 1,4-divinyl naphthalene. The formation of (*M*)-(–)-**133** involves a sequential process consisting of the cycloaddition of **112**, pyrolytic sulfenic acid elimination, and final aromatization, in each of the two dienic moieties of the starting naphthalene.

Taking advantage of the high stereoselectivity of the Diels-Alder reactions of sulfinyl quinones with acyclic dienes, the synthesis of the (+)-royleanone [118c] has been carried out. The reaction of 2-hydroxy, 3-*i*-propyl, 6-*t*-butylsulfinyl 1,4-benzoquinone with 1,3,3-trimethyl, 2-vinyl cyclohexene under high pressure (under thermal or catalytic conditions the reactions do not work) yielded optically pure intermediate A (Scheme 66), resulting from cycloaddition followed by elimination of the sulfinyl group. This result suggests that cycloaddition has taken place with total control of the regio, *endo*, and π-facial selectivities. Starting from similar quinones containing *p*-tolylsulfinyl groups, the optical purity of the intermediates such as A are high (90–92% ee) but not complete. Intermediate A was further hydrogenated with many different reducing agents, yielding mixtures of (+)-royleanone and its epimer (Scheme 66). Hydrogenation with Pd/C yielded a 60:40 mixture of the two possible diastereoisomers. The major one was easily isolated by chromatography and further identified as optically pure (+)-royleanone (Scheme 66).

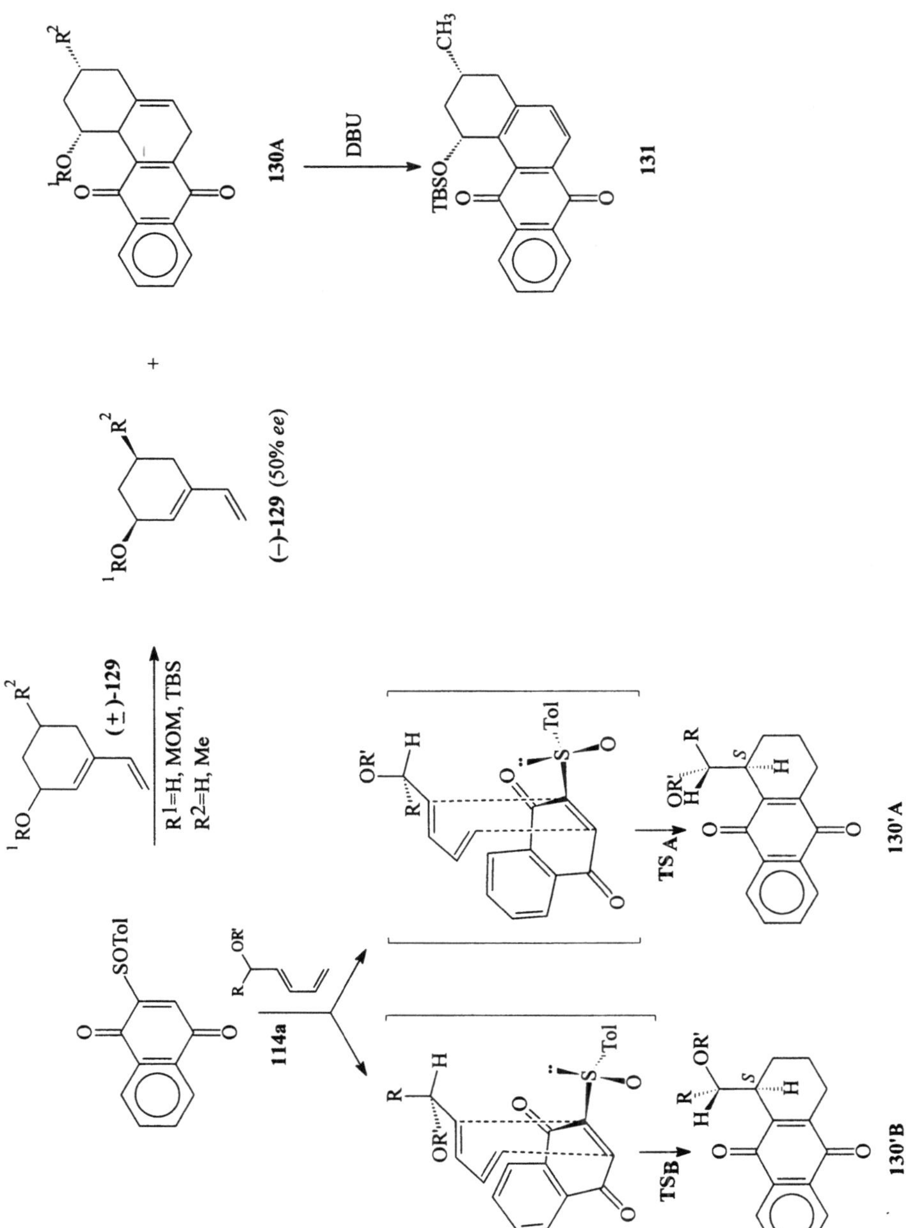
130A
DBU
131
(−)-129 (50% ee)
(±)-129
R1=H, MOM, TBS
R2=H, Me
114a
TS A
TS B
130'A
130'B

Scheme 65

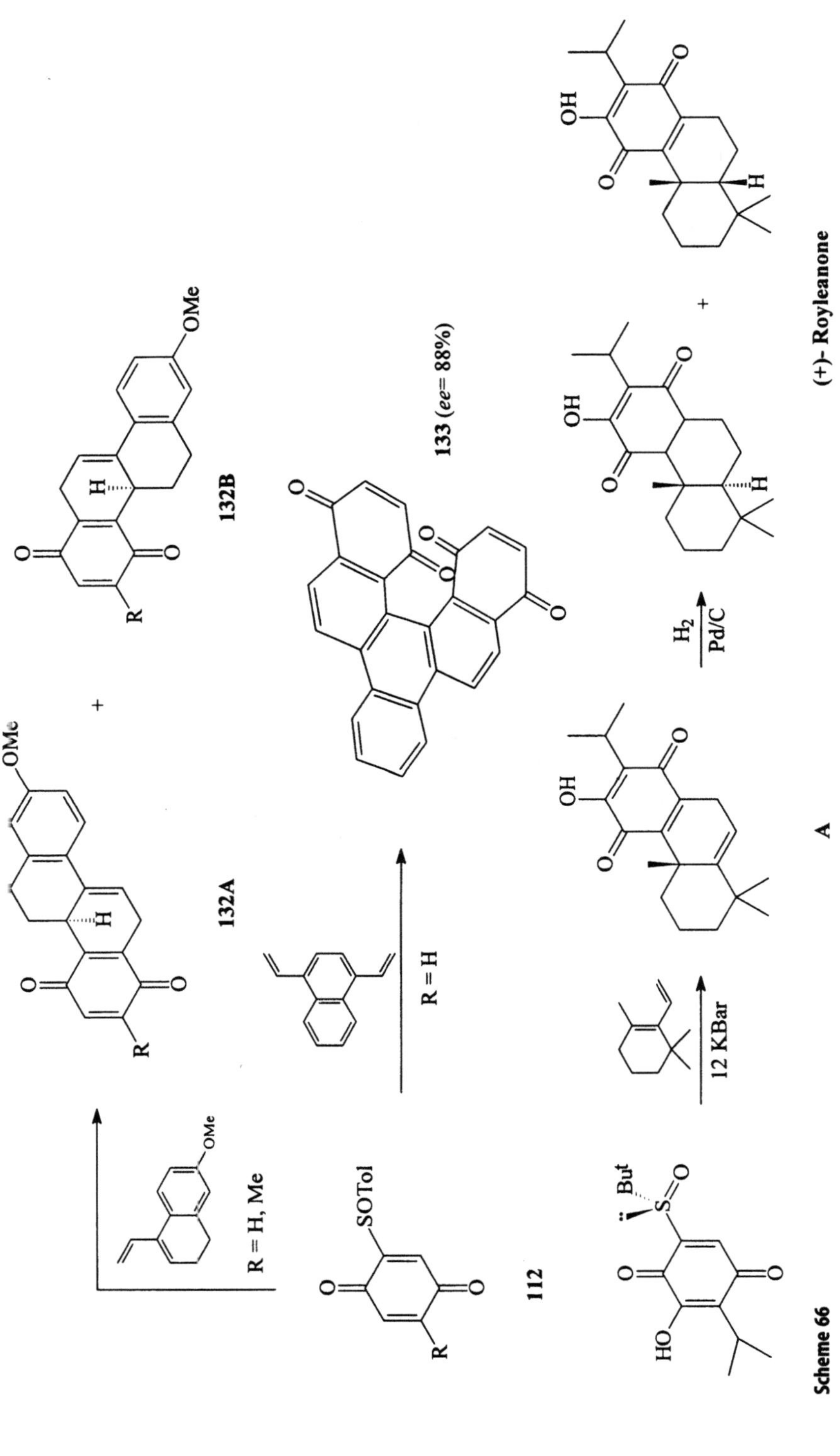

R = H, Me
SOTol
112
132A
132B
OMe
R = H
133 (ee= 88%)
12 KBar
H2
Pd/C
A
(+)- Royleanone

Scheme 66

2.2
Sulfinyl Dienes

As we have shown, the sulfinyl group has been widely used as a chiral inductor in Diels-Alder reactions when bonded to the dienophilic double bond, due to its strong ability to control the π-facial selectivity. However, only a small number of papers on the use of sulfinyl dienes in asymmetric synthesis have been written, perhaps due to the poor reactivity of many of these substrates and the complex course of their reactions (mainly in the case of 1-sulfinyl dienes, see later). Additionally, the fact that synthetic methods to obtain enantiomerically pure sulfinyl dienes have been available only in the last six years would also explain the low number of papers concerning these asymmetric Diels-Alder reactions. During the preparation of this account, an excellent review on the synthesis and asymmetric Diels-Alder reactions of chiral 1,3-sulfinyl dienes has been published [11].

A number of different methods have been used to synthesize racemic sulfinyl dienes. They involve oxidation of dienylthioethers [119], reactions of α-sulfinyl carbanions with different electrophiles [60, 120], sulfoxide-sulfenate rearrangements of propargylic sulfenates followed by isomerization of the resulting sulfinylallenes [121], and reactions of sulfolenes with Grignard reagents [122].

The synthetic strategies used to prepare optically active sulfinyl dienes (Schemes 67 and 68) differ depending upon the position (C-1 or C-2) of the sulfinyl group at the diene. In addition to oxidation of chiral thioethers into the cor-

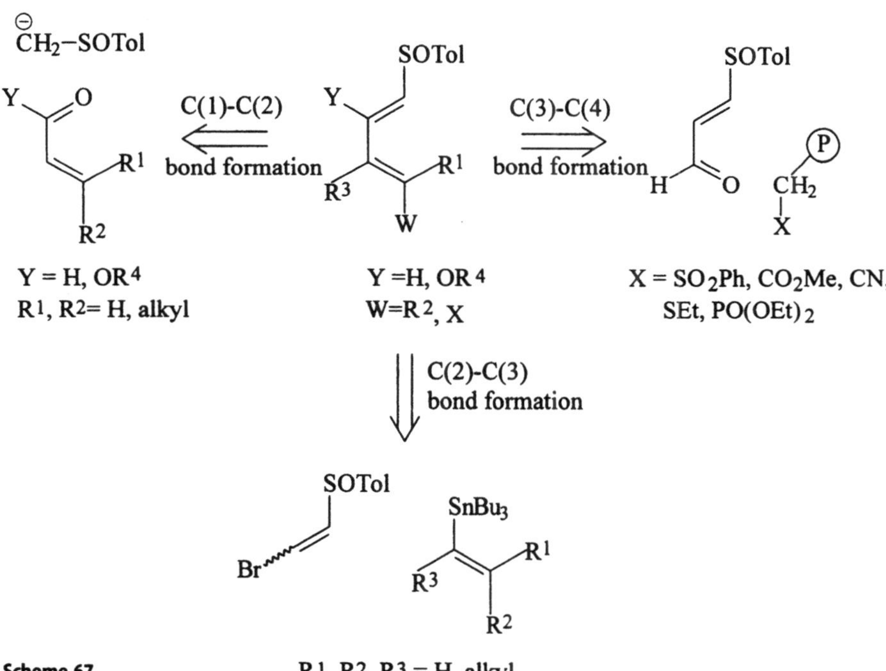

Scheme 67

responding sulfoxides [123], three general methods have been recently developed to obtain 1-sulfinyl dienes (Scheme 67). The first involves the formation of the C(1)-C(2) bond of the butadiene skeleton by condensation of α-sulfinyl carbanions with appropriate α,β-unsaturated carbonyl compounds [124]. Palladium-catalyzed coupling of enantiopure 2-halovinyl sulfoxides with (*E*) or (*Z*)-vinyl (or alkinyl) stannanes has been used to create the C(2)-C(3) bond of dienic systems [125] and the Horner-Wadsworth-Emmons reaction of (*R*)-3-*p*-tolylsulfinyl propenal with appropriate phosphonates afforded 4-X, 1-sulfinyl dienes (X = SO_2Ph, CO_2Me, $PO(OEt)_2$, CN, SEt) by formation of the C(3)-C(4) bond [125a, 126].

Several different methods have been used to synthesize non-racemic 2-sulfinyl dienes (Scheme 68). Those involving the formation of the C-S bond, by addition of chiral sulfenic acids to enynes [127] (route a, Scheme 68) or rearrangement of conjugated propargylic sulfenates [128], afforded mixtures of sulfoxides epimeric at sulfur, which were easily separated in most cases. Other methods avoid this problem by starting from optically pure vinyl sulfoxides and creating the C(2)-C(3) bond of the dienic system. To this end, palladium-catalyzed coupling of 1-halovinyl sulfoxides with vinylstannanes [129] are very efficient (route b, Scheme 68), whereas those methods involving transformation of the condensation products of vinyl *p*-tolyl sulfoxide and aldehydes, ketones, or esters [130] are quite versatile but their yields are usually poor (route c, Sche-

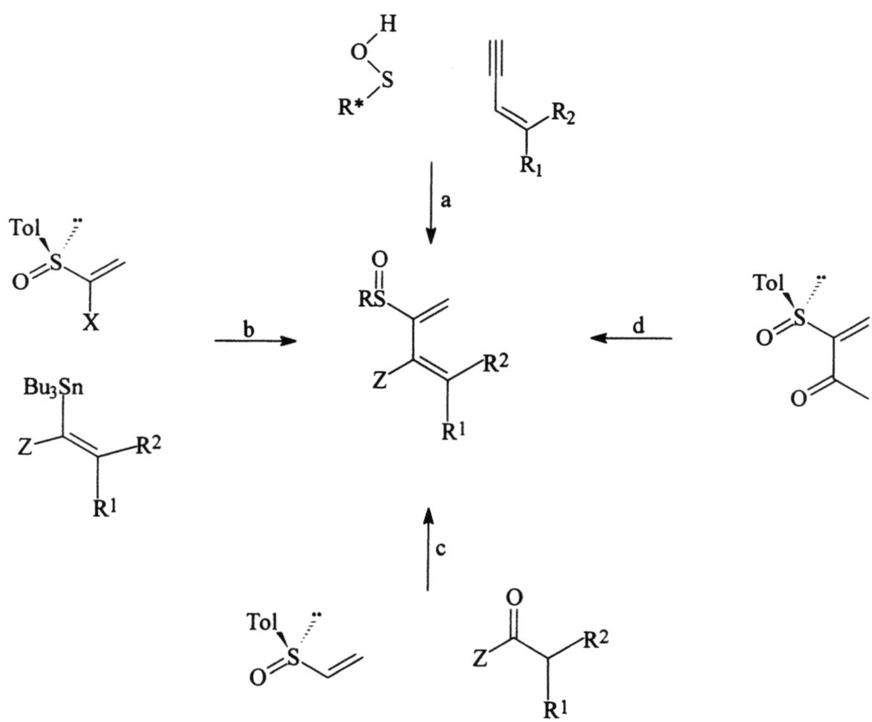

Scheme 68 Z = H, alkyl, OR

me 68). The use of optically pure α-methylene β-ketosulfoxides (obtained by Mannich reaction of β-ketosulfoxides) as starting materials to prepare 3-alkoxy 2-*p*-tolylsulfinyl butadienes (route d, Scheme 68) has been described recently [131]. A thorough description of all these methods can be found in the review of Aversa et al. [11].

Taking into account that a significant part of the current knowledge of the stereochemical course of Diels-Alder reactions of sulfinyl dienes has been deduced from the studies made on racemic substrates, a detailed consideration is advisable herein. The first cycloaddition of a racemic sulfinyl diene was reported in 1972 by Evans et al. [119a]. The reaction of (±)-1-butadienyl phenyl sulfoxide **134** with the tetrasubstituted enamine **135** required 24 h at 70 °C to afford a diastereoisomeric mixture of the *endo* adducts **136** (Scheme 69). A small amount of aminoalcohol **137** was also obtained. The latter compound resulted from the sulfoxide-sulfenate rearrangement of **136**, and hence it was the only product obtained after treatment of the crude reaction mixture with $Na_2S \cdot 9H_2O$, which favors the formation of the alcohol in such rearrangements. The purpose of this paper was to define the complementary nature of certain [4+2]-cycloaddition and [2, 3] sigmatropic processes in order to obtain substituted cyclohexene derivatives, which are quite difficult to obtain using the direct Diels-Alder routes.

From the unequivocally established *syn* relationship between hydroxy and amino groups in **137**, the same relative stereochemistry of the sulfoxide and amino functions in the precursor adducts **136** may be inferred based upon the stereospecific character of the rearrangement. On this basis, it could be concluded that Diels-Alder reaction of sulfinyldiene **134** took place with complete control of *endo* and regioselectivities.

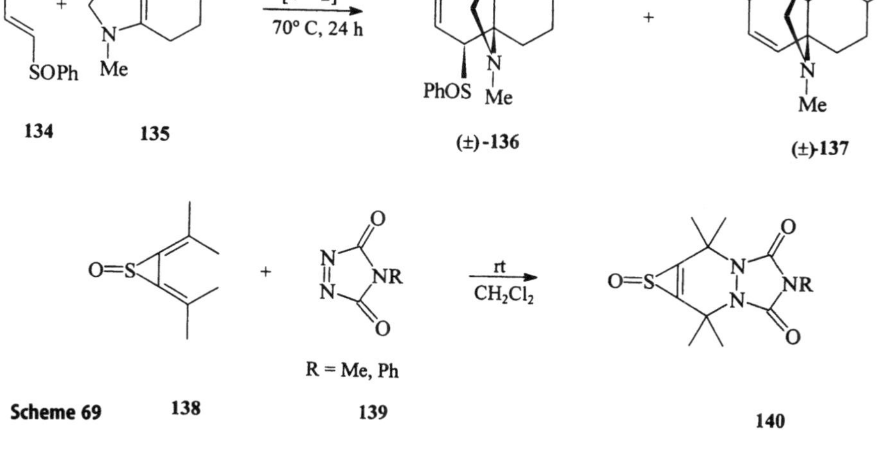

134 **135** **(±)-136** **(±)-137**

Scheme 69 **138** **139** **140**

Ten years later, the synthesis of alkyl-substituted thiirene sulfoxide **140** was reported (Scheme 69), making use of a Diels-Alder reaction between thiiranoradialene sulfoxide **138** and the highly reactive dienophilic 4-substituted 1,2,4-triazoline-3,5-diones **139** (TAD) [132]. The fact that other classical reactive dienophiles, including maleic anhydride and ethyl azodicarboxylate, do not react with sulfinyldiene **138**, reveals its low reactivity.

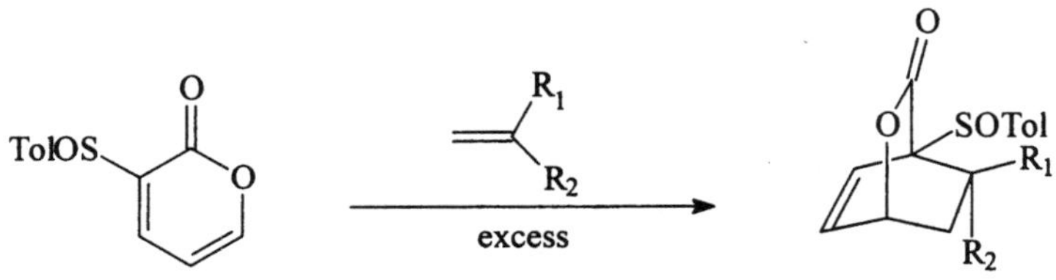

(±)-**141** (±)-**142**

R_1	R_2	conditions	yield (%)	*endo*:*exo*	*de** (%)
OMe	OMe	rt, 2d	95	100:0	76
H	OEt	$ZnBr_2$, rt, 3d	87	10:1	60
H	SMe	6.8 Kbar, rt, 24h	98	100:0	100
H	SPh	6.8 Kbar, rt, 3d	73	100:0	94

* of the major *endo* adducts

Scheme 70

Subsequently, Posner published the completely regioselective and highly stereoselective cycloadditions of racemic 3-(*p*-tolylsulfinyl)-2-pyrone (**141**) (Scheme 70) with 1,1-dimethoxyethylene [133], vinylether, and vinylthioethers [134]. With the first dienophile, the best diastereoselectivity (an 88:12 ratio of the two *endo*-adducts) was achieved at room temperature in toluene or hexane as the solvent (48 h). A 10:1 *endo/exo* mixture of cycloadducts was obtained with vinylether in the presence of $ZnBr_2$ as the catalyst, whereas a total *endo* selectivity was observed in reactions of **141** with vinylthioethers [134] conducted under high pressures. The bridged bicyclic lactone cycloadducts **142** have been used as versatile synthons in the synthesis of shikimic acid derivatives. Although enantiomerically pure samples of compound **141** could be obtained [134] it was not used as a starting material for asymmetric Diels-Alder reactions (the low yield of (*S*)-**141** precluded this).

Reactions of acyclic 1-phenylsulfinyl-4-(alkoxycarbonyl)amino butadienes **143** with phenyl vinyl ketone and acrolein were studied by Overman et al. (Scheme 71) [119b]. Good reactivity and complete regioselectivity (controlled by the nitrogen moiety) were the main characteristics exhibited by these dienes. Additionally, the *endo* approaches are clearly favored with respect to the *exo* ones (*endo/exo* ratio is 10 for aldehyde and higher than 100 for ketone). Unfortunately, the π-facial selectivity was small or non-existent in both *endo* and *exo* approaches, yielding a 1:1 mixture of the two possible diastereoisomers (Scheme 71).

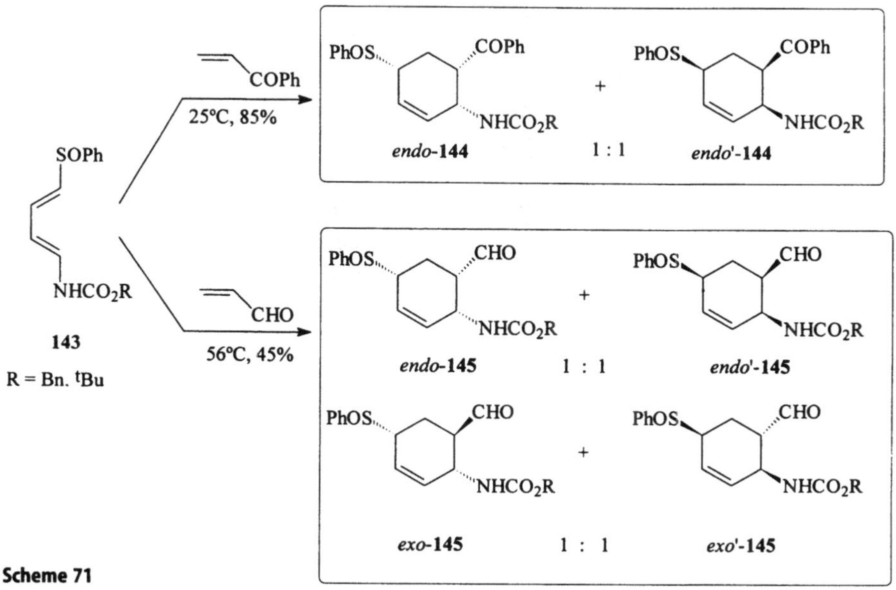

Scheme 71

Based on the results obtained in the reactions of sulfinyl dienes **146** [119d] and **147** [119d, 120a] (Scheme 72) with *N*-phenylmaleimide (NPM), Overman et al. proposed the first stereochemical model for these cycloadditions [119d]. These dienes reacted with a total control of the *endo* and π-facial selectivities, the approach of the dienophile toward the face of diene opposite to that containing the sulfinyl oxygen (the upper face in Scheme 72) being favored. This was explained by assuming a strong electrostatic repulsion between the sulfinyl oxygen at the diene and the carbonyl oxygens at the dienophile, which would destabilize the transition states corresponding to the approaches toward the face containing the sulfinyl oxygen.

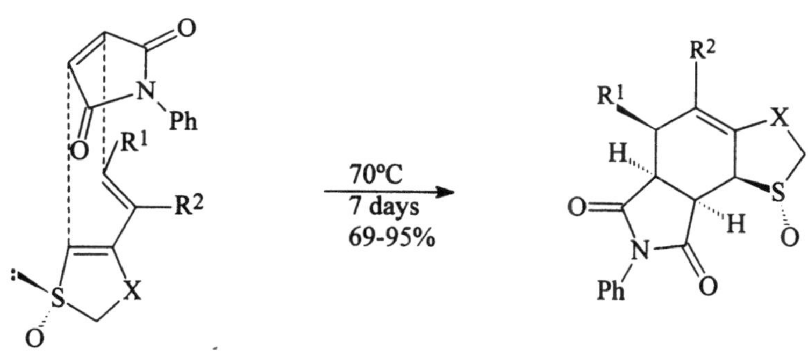

146 ($X = CH_2$)
147 ($X = O$)

Scheme 72

In a further study, Fallis et al. [119e,f] proposed that cycloadditions occurred exclusively in a contrasteric manner *syn* to the sulfur-oxygen bond, which is the opposite to that postulated by Overman et al. [119d]. This proposal is based on the results obtained in reactions of 2,5-dimethylthiophene-*S*-oxide **148** (generated in situ by MCPBA oxidation of 2,5-dimethylthiophene) with different electron poor dienophiles, including quinones, NPM, tetracyanoethylene, and 2-chloroacrylonitrile, which took place with complete *endo* and π-facial selectivities (with other classical dienophiles the adducts could not be isolated) and gave the adducts resulting from the approach of the dienophile to the face of the diene containing the sulfinyl oxygen (Scheme 73). In order to explain these results, which are contradictory to those observed by Overman, Fallis invoked Cieplak's model [135], and argued that the ability of the lone electron pair at sulfur to stabilize the developing incipient σ-bond, is stronger than that of the sulfinyl oxygen, determining that this contrasteric course was favored for the diene approach [119e,f]. This special behavior seem to be characteristic of thiophene *S*-oxides and it has been observed in its reactions with other dienophiles [136].

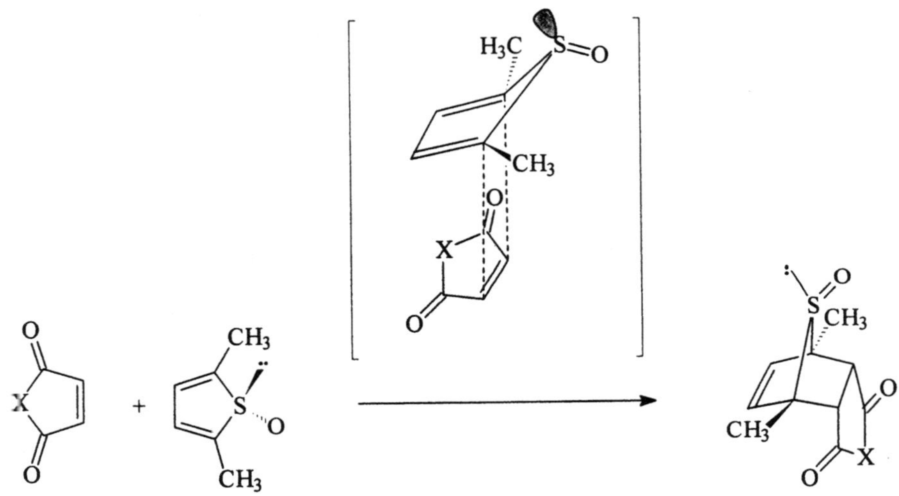

Scheme 73 **148**

1.2.1
Optically Pure 2-Sulfinyldienes

In 1988 Gibbs and Okamura [128] described the intramolecular Diels-Alder reaction of the non-racemic vinylallene **150** to afford the adduct **151** (as a mixture of epimers at sulfur) in a completely *exo*-selective manner, due to the topographical and steric arrangement of the starting vinylallene (Scheme 74). Compound **150** was obtained as a mixture of epimers at sulfur from the optically pure propargylic alcohol **149** (this transformation involved a sulfoxide-sulfenate rearrangement). Compound **151** was used to synthesize (+)-sterpurene.

After this particular and specific reaction, which may be considered the first asymmetric Diels-Alder reaction using optically active sulfinyldienes, further examples of reactions with enantiomerically pure sulfinyldienes were not reported until 1993. Aversa et al. [137] accomplished a study of the reaction of (R_S) and (S_S)-*E*-3-[(1*S*)-isobornyl-10-sulfinyl]-1-methoxy-1,3-butadiene (**152a**) with methyl acrylate in the presence of different Lewis acids (in Scheme 75 only the structures of compounds derived from (R_S)-dienes are depicted). The combined directing effects of the substituents at the 1- and 3-positions in these dienes ensured that cycloaddition occurred with complete regioselectivity. An almost complete *endo* selectivity as well as a very high π-facial selectivity, (*de* >90%) for the *endo* approach, were observed in reactions catalyzed by $LiClO_4$, which gave a 96:4 mixture of *endo* **A-153a** and *endo* **B-153a** (Scheme 75). Similar stereochemical results but very poor yields were obtained in the presence of $ZnCl_2$. The absolute configuration of the major adduct (*endo* **A-153a**) was unequivocally established as that depicted in Scheme 75 by X-ray diffraction studies. In the absence of catalysts, a 54:31 mixture of the same adducts was obtained (the formation of the *exo* adducts could also be detected). An opposite π-facial selectivity was observed when diene (S_S)-**152** was used as the starting material. These facts demonstrate that the sulfinyl group seemingly is the main controller of the stereochemical course of the cycloaddition. This was rationalized for catalyzed reactions in terms of the mutual coordination of the metal with sulfinyl oxygen and carbonyl oxygen in transition state structures such as **A** (Fig. 13),

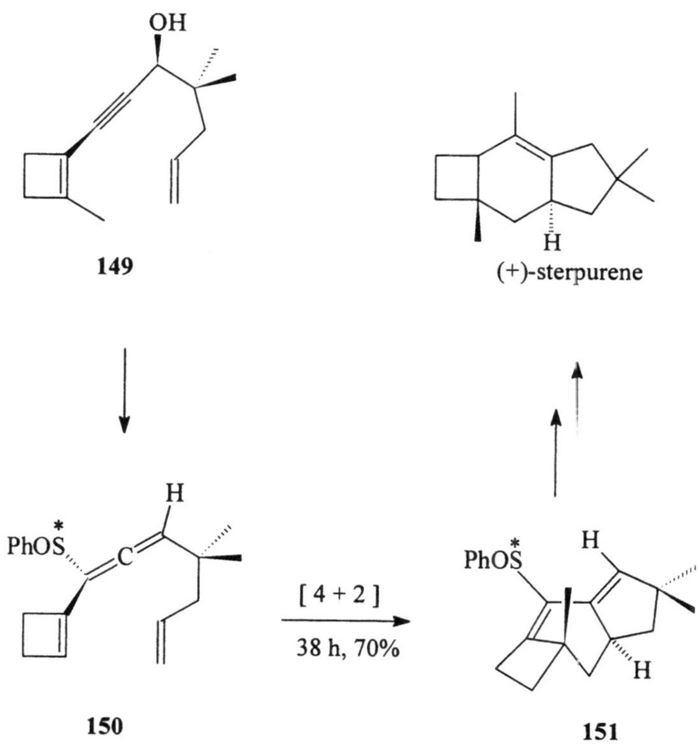

Scheme 74

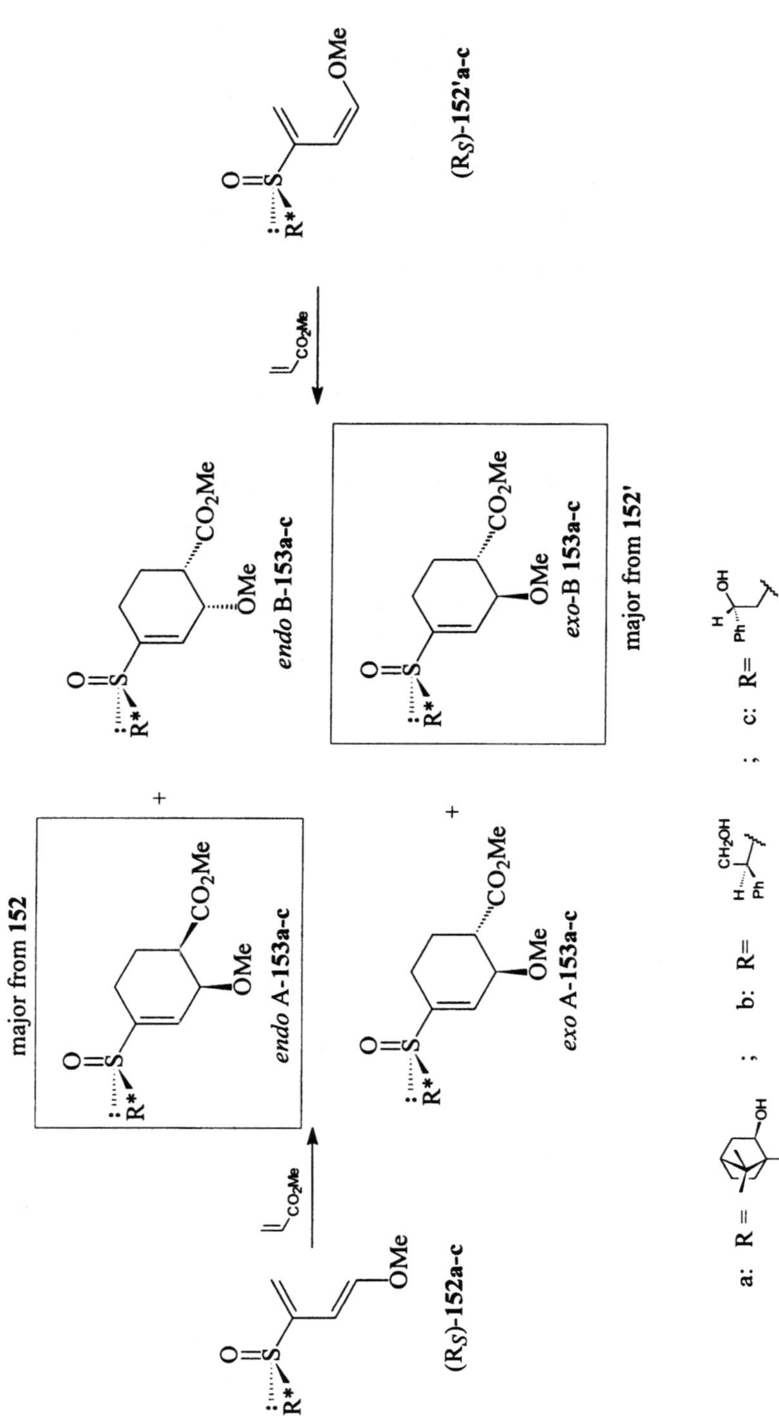
major from 152
endo A-153a-c
endo B-153a-c
exo A-153a-c
exo-B 153a-c
major from 152'
(R$_S$)-152a-c
(R$_S$)-152'a-c
a: R =
b: R=
c: R=

Scheme 75

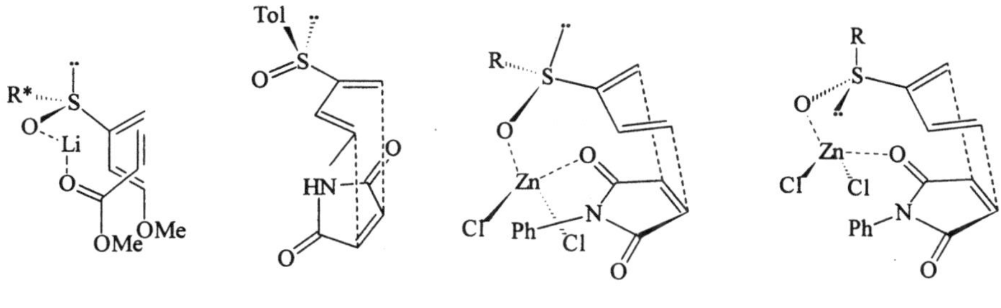

Fig. 13

with the isobornyl group subject to minimal steric compression [138]. No explanation of the stereochemical course of these reactions in the absence of catalysts (lower selectivity but predominance of the same *endo* **A-153a** adduct in the reaction mixture) was formulated.

In a later paper, the authors comment on the difficulties they found in the removal of the chiral auxiliary under mild conditions due to the presence of the mercaptoisoborneol moiety. Moreover they report that the use of other sulfinyldienes **152b** and **152c** (Scheme 75), derived from 2-hydroxy-1-phenylethanothiol and 2-hydroxy-2-phenylethanothiol respectively, gave stereochemical results similar to those of **152a**, and have certain advantages related to the easier preparation and elimination of these sulfur residues [127b]. Recently, Aversa et al. [137b] have reported the full results of their studies on the reactions of different (*E*) and (*Z*)-3-alkylsulfinyl-1-methoxy-1,3-butadienes **152a–c** and **152′a–c** with methyl acrylate. The influence of the chiral auxiliary (R groups) and the stereochemistry of the OMe group on the reaction course is evaluated. All these reactions proceeded with complete regioselectivity and very high stereoselectivity when catalysed by $LiClO_4$ or $ZnCl_2$ in CH_2Cl_2, regardless of the chiral auxiliary used, thus demonstrating that chirality at sulfur is the main controller of diastereofacial selectivity. Compounds *endo* **A-153** and *exo* **B-153** were obtained as the major products from (R_S)-**152** and (R_S)-**152′** respectively (Scheme 75). Reactions of the (*Z*)-isomer **152′** were much more sluggish.

Further similar stereochemical results were reported by Maignan et al. [139] concerning the reactions of (*E*)-(+)-(*R*)-2-*p*-tolylsulfinyl-1,3-pentadiene (R_S)-**154**) with maleimide. The adduct **155** (Scheme 76) was exclusively obtained at room temperature (10 h without catalyst) and its structure was assigned by X-ray diffraction studies.

In order to justify the stereochemistry of the major adducts **155**, the authors proposed that they result from an exclusive *endo* approach toward the less hindered face of the diene (opposite to the *p*-tolyl group), with the sulfinyl group adopting the s-*trans* conformation with respect to the C(1)-C(2) double bond (**B** in Fig. 13). Nevertheless, this conformational preference is not rationalized in the paper. Aversa et al. [137c] have recently reported that uncatalyzed reactions of (R_S)-**152** with maleimide and *N*-phenyl maleimide occurred with complete *endo*- and very high facial diastereoselectivities. In this sense, the results ob-

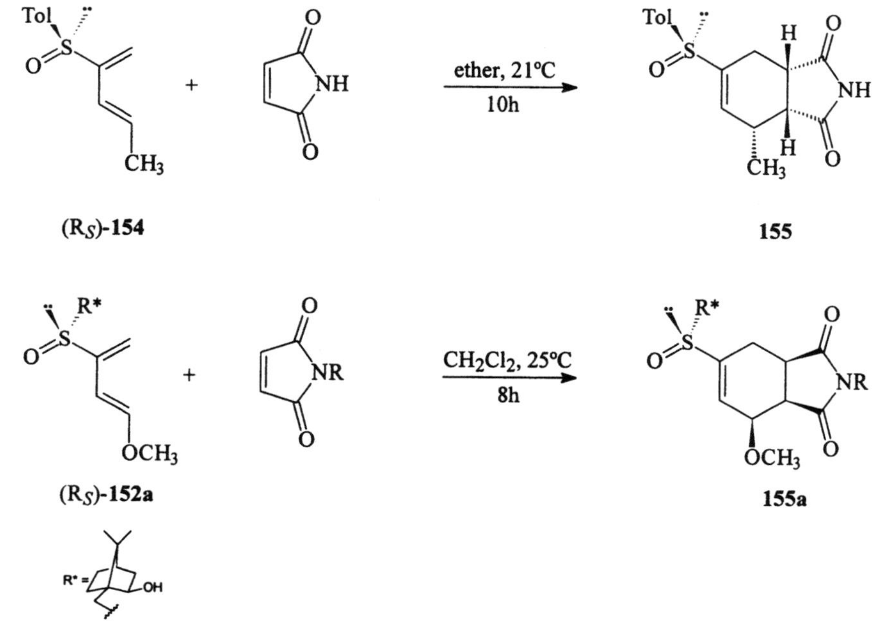

Scheme 76

tained with maleimide affording **155a** (see Scheme 76), are identical to those of (R_S)-**154**. The fact that dienophile approach takes place from the upper face of **152**, whereas it occurs from the bottom face of **154**, can be explained by taking into account that the spatial arrangement of the substituents around the sulfur is opposite in **152** and **154**.

Sulfinyl diene (R_S)-**156**, derived from (R_1,S_2,R_3) -3-mercaptocamphan-2-ol, reacts with *N*-phenylmaleimide [123b] to afford mixtures of two *endo* adducts **157a** and **157b** (Scheme 77), with the second as the major product. The π-facial selectivity is reversed and become almost complete in the presence of $LiClO_4$ and X_2Zn.

The stereochemistry of the adducts **157B** (Scheme 77) and *endo* **A-153** (Scheme 75), obtained as the major ones in the non-catalyzed reactions of (R_S)-**156** and (R_S)-**152** respectively, suggests that they must be formed through a similar stereochemical course. This was explained [123b] by assuming the model proposed by Aversa et al. [137a] (electrophilic *N*-phenyl maleimide prefers to add to the electron-rich face of the nucleophilic diene). In the presence of $LiClO_4$ and $ZnCl_2$, the sense of the π-facial selectivity was inverted for **156** but was retained for **152**, revealing significant differences in their respective stereochemical courses. According to the authors, the presence of the Lewis acids induces the formation of species **C** (Fig. 13), which gradually turned the stereoselectivity around depending upon the size of the catalyst. Nevertheless, taking into account the *s-trans* arrangement of the sulfinyl oxygen with respect to the C(1)-C(2) double bond in species **C**, its association with the carbonyl oxygen through the catalyst seems unlikely.

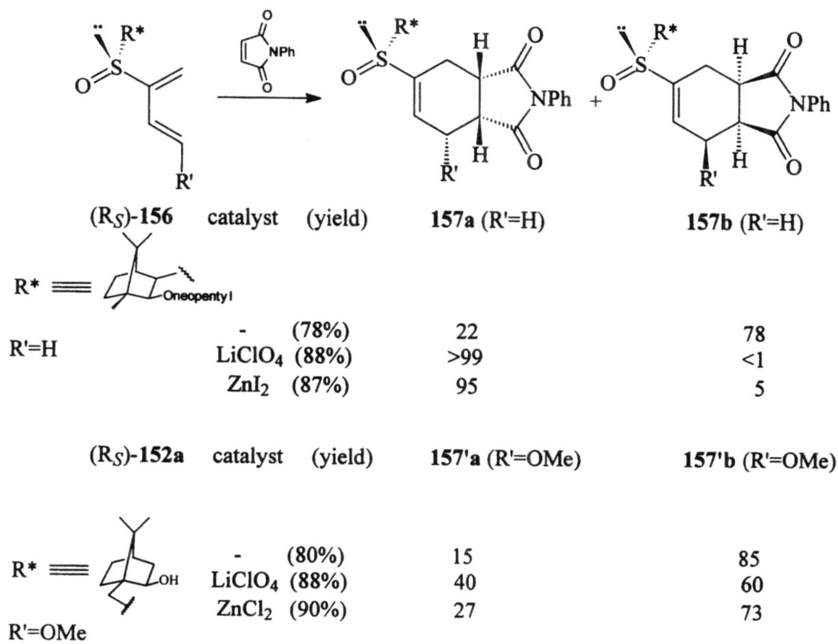

Scheme 77

The formation of chelated species such as **D** (Fig. 13) would better explain the stereochemical results.

The reactions of (R_S)-**152a** with *N*-phenyl maleimide (Scheme 77) affords a 15:85 mixture of **157′a** and **157′b**. The stereoselectivity decreases when $LiClO_4$ or $ZnCl_2$ are used as catalysts [137c]. This behavior contrasts with that reported for (R_S)-**156** (see Scheme 77), which afforded almost exclusively **157a** in the presence of these Lewis acids, but this is not discussed in the paper. Otherwise the reactions with diethyl maleate are completely stereoselective under $LiClO_4$ catalysis, although a mixture was obtained under thermal conditions [137c]. Two different stereochemical models are suggested to explain the evolution of these reactions with cyclic and acyclic dienophiles.

Pradilla et al. [140] have recently produced a nice paper showing that enantiopure hydroxy 2-*p*-tolylsulfinyl butadienes **158** (Scheme 78) undergo a highly face-selective Diels-Alder cycloaddition with *N*-phenyl maleimide and phenyltriazolidine dione, presumably controlled by the chiral sulfur atom (dienophile approach from the upper face of diene). Complementary π-facial selectivity (dienophile approach from the bottom face of diene) is displayed by related enantiopure sulfonyldienes **158′** (Scheme 78). The authors suggest that the behavior of **158** is a consequence of the predominant influence of the chiral sulfur with respect to the hydroxylic carbon (the only chiral center in **158′**) on the stereochemical course of the cycloadditions. According to their explanation, dienes will adopt conformations similar to those depicted in Scheme 78, with the chiral centers employing their stereochemistry to maximum effect due to 1,3-allylic strain (which is considered as the main directing effect of these cycload-

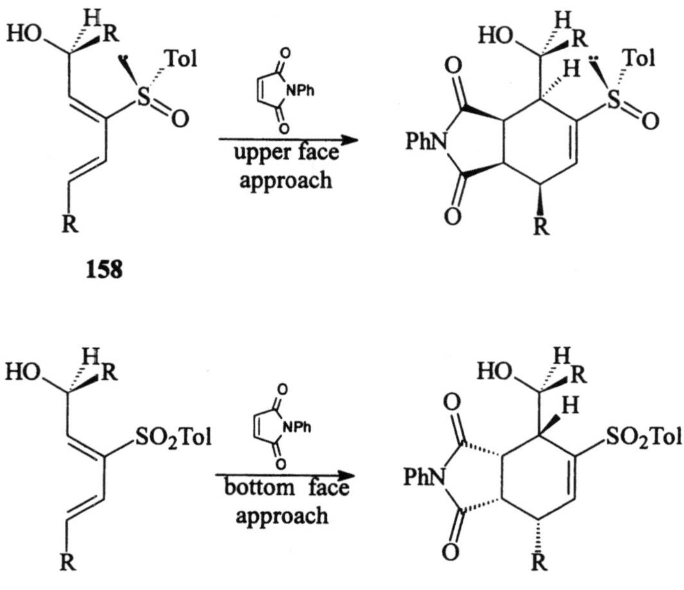

158

158'

Scheme 78

ditions). As the approach of diene to the upper face of **158** is clearly favored, the authors deduce the predominant role of the sulfinyl group. By contrast, the preferred attack takes place to the bottom face of **158′** because the hydroxy bearing carbon is the only controller of the dienophile approach. Other factors, including the possible influence of hydrogen bonding and the substituent at nitrogen, which could throw some light on the origin of the stereoselectivity, have not been investigated.

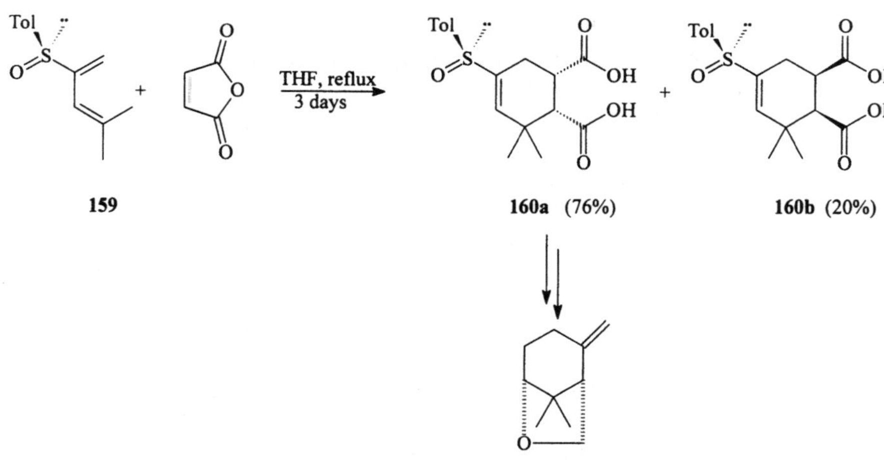

159 **160a** (76%) **160b** (20%)

(-)-Karahana ether

Scheme 79

The first application of the asymmetric Diels-Alder reaction of 2-sulfinyldienes in the field of natural products chemistry was the synthesis of (–)-(1*S*,5*R*)-karahana ether [141]. The key step of this synthesis involved the reaction of (+)-(*R*)-4-methyl-2-(*p*-tolylsulfinyl)-1,3-pentadiene (**159**) with maleic anhydride, which afforded an 80:20 mixture of the two carboxylic acids **160a** and **160b**, presumably resulting from the hydrolysis of the two bicyclic anhydrides obtained by *endo*-approach of dienophile toward both diastereotopic faces of the diene. The structure of the major component of the mixture (which was used as the starting material in the synthesis of the Karahana ether) was tentatively assigned to **160a** (Scheme 79) on the basis of previous results concerning Diels-Alder reactions on sulfinyldiene **154** [139]. However, the reasons explaining that the lower π-facial selectivity observed in reactions of **159** compared with that of **154** are not discussed in the paper. We must remark that **159** and **156** exhibit a similar π-facial selectivity in their respective reactions with maleic anhydride and *N*-phenylmaleimide in the absence of catalyst (compare Schemes 79 and 77, taking into account the different spatial arrangement of the substituents around the sulfur in both dienes).

The available results on the stereochemical behavior of 2-sulfinyl dienes can be rationalized by assuming that the favoured approach of dienophile takes place toward the less hindered face of diene, which adopts a conformation minimizing the electrostatic repulsion of the sulfinyl oxygen and the heteroatoms at the dienophile. Thus, the attack of maleic anhydride or maleimides at the bottom face of 2-sulfinyl dienes (with the configuration shown in Fig. 14) in an *s-trans* conformation will be preferred (**A** in Fig. 14). Steric and electrostatic interactions between the sulfinyl oxygen and X grouping would explain the π-facial selectivity observed in the reaction conducted in the absence of catalysts (the proportion of diastereoisomers are indicated in brackets in Fig. 14). The same transition state would be unstabilized in reactions of **152** with acrylates (**B′** in Fig. 14) due to electrostatic repulsion and other conformations minimizing it, such as those shown in TS **B** and **C** (each one favoring the approach of dienophile to opposite faces).

In the presence of a catalyst able to become associated with the basic centers of the reagents, conformational preferences of dienes change to allow such an association. Hence, reactions of **156** with *N*-phenyl maleimide can be explained by assuming a TS of the form **D** (Fig. 14), where the sulfinyl oxygen adopts the *s-cis* arrangement. This explains the inversion of the π-facial selectivity with respect to that observed in the absence of a catalyst. Finally, reactions of methyl acrylate with diene **152** can be satisfactorily explained from the TS proposed by Aversa et al. [137c] (**A** in Fig. 13), or TS **E** in Fig. 14, which is similar but displays the sulfinyl oxygen in an *s-trans* arrangement.

2.2.2
Optically Pure 1-Sulfiny Dienes

Asymmetric Diels-Alder reactions of 1-*p*-tolylsulfinyldienes **161** with *N*-methylmaleimide (NMM) were completely stereoselective both under thermal and catalyzed conditions [142], yielding the same compounds *endo*-**162** as sole

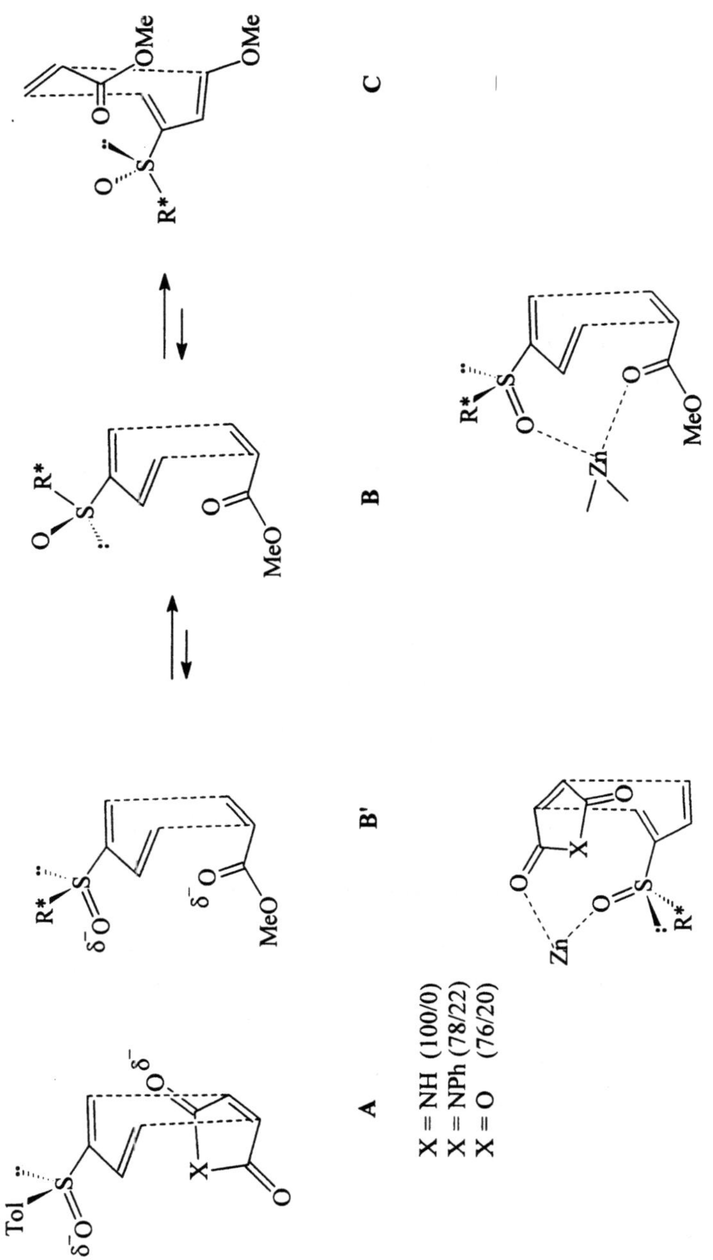

Fig. 14

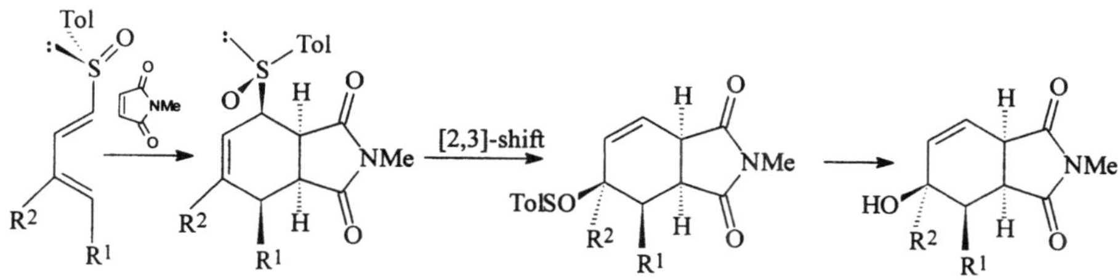

161a (R^1=Me, R^2=H)
161b (R^1= Ph, R^2=H)
161c (R^1=OEt, R^2=CH_3)

162 (a-c)

163 (a-c)

Scheme 80

adducts. The reactivity of the diene is very low, even in the presence of Lewis acids acting as catalysts (more than 20 days are required for these reactions to complete), and the π-facial selectivity is completely controlled by the sulfinyl group. Electron-rich dienophiles, such as ethyl vinyl ether and 3,4-dihydro-2*H*-pyran yielded no cycloadducts either thermally or under Lewis acid catalyzed conditions. Moreover, the reactions of **161a** with other electron-poor dienophiles different from maleimides do not occur (this is the case with methyl propiolate and ethyl acrylate) or they do so only very slowly (after one month, reaction with p-benzoquinone affords only traces of aromatized Diels-Alder adduct). All these facts point to the low reactivity of these 1-sulfinyl dienes.

Under catalytic conditions, adducts **162a–c** were isolated in high yields (Scheme 80). However, in the absence of catalysts, the initially formed adducts **162** were transformed in situ into enantiomerically pure highly functionalized cyclohexenol derivatives **163**, which are not accessible by direct cycloaddition, through the stereospecific sulfoxide-sulfenate rearrangement followed by desulfinylation (it seems that NMM acts as a thiophilic agent). The stereochemical course of the cycloaddition was explained by considering the relative stability of the transition states resulting from the *endo* approach of dienophile toward the less hindered face of the possible conformations of the diene around the C-S bond. The minimum steric and electrostatic repulsions between the carbonyl

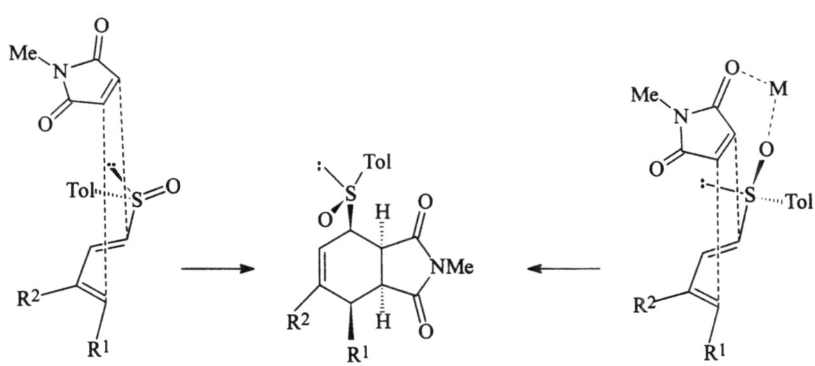

Fig. 15 A B

oxygen of NMM and the sulfinyl oxygen must be those existing in the TS **A** indicated in Fig. 15, with the diene adopting the *s-trans* disposition of the S=O and C(1)=C(2) bonds. A similar explanation was postulated to explain the behavior of the 2-sulfinyldienes in the absence of Lewis acids (see above). A chelated species **B** (Fig. 15), involving coordination of metal **M** with sulfinyl and carbonyl oxygens, could explain the stereochemical results observed in the presence of the Lewis acids.

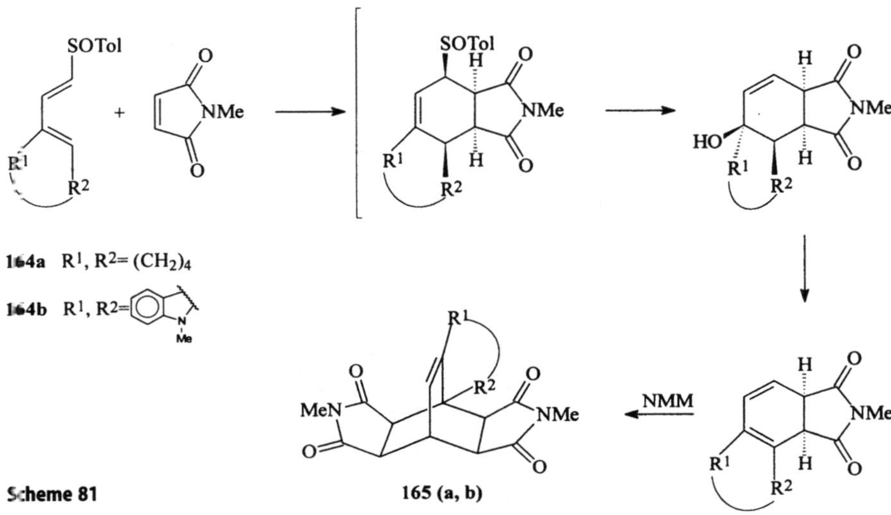

Scheme 81

(*R*)-(1*E*,3*E*)-2-Cyclohexenyl-1-vinyl *p*-tolylsulfoxide (**164a**) and (*R*)-(1*E*,3*E*)-*N*-methyl-3-[2-(*p*-tolylsulfinyl)vinyl]-1*H*-indole (**164b**) reacted with *N*-methylmaleimide, giving rise stereoselectively to highly substituted bycyclo [2,2,2] octenes **165a** and **165b** in a one-pot four-step procedure through a tandem Diels-Alder reaction/sulfoxide-sulfenate rearrangement/dehydration/[4+2]-cycloaddition sequence [124e] (Scheme 81). A similar process was observed with *N*-phenyl maleimides [124f].

The evolution of the adduct obtained by reaction of maleic anhydride with sulfinyldiene **161a** was unexpected and very complex [143a]. When the reaction was accomplished under thermal conditions at normal pressure, a mixture of lactones **167** and **168** was isolated after several days of reaction, their relative proportion ranging between 90:10 and 70:30, depending on the temperature and the solvent polarity (Scheme 82). The initial adduct **166** was not observed under these conditions. Under high pressures, a 70:30 mixture of the initial adduct **166** and lactone **167** was obtained after several hours, which slowly evolved (ten days) at room temperature into the same mixture of lactones obtained under normal pressure. The optical purity of the major lactones depends on the conditions used, being the highest (*ee*=82%) when reactions were carried out under high pressures.

The formation of lactone **167** as the major compound derived from **166** was rationalized as follows: the intramolecular acylation of the sulfinyl oxygen

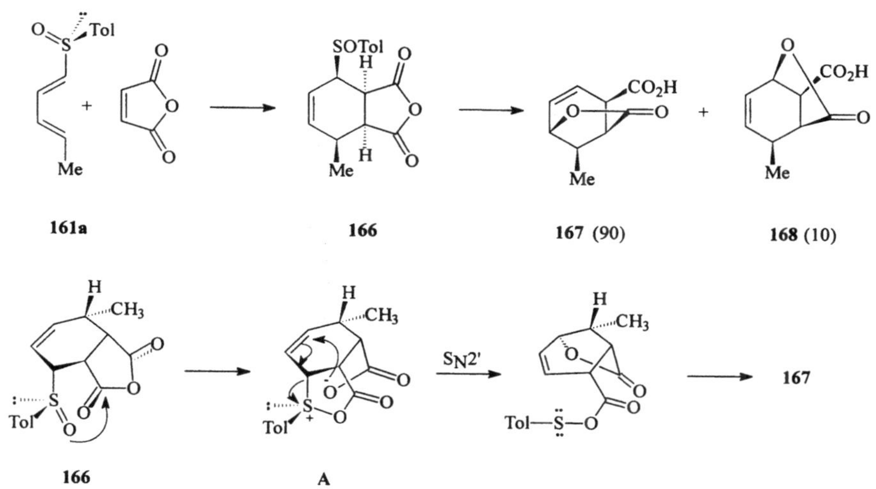

Scheme 82

would yield compound **A** (Scheme 82), which proceeds through an S_N2' process (intramolecular attack of the free carboxylate onto the allylic acyloxy sulfonium with elimination of the sulfur function) and further hydrolysis of the resulting sulfinate to afford compound **167** (Scheme 82). Hydrolytic opening of the intermediate **A** by traces of water (present in the reaction medium when long reaction times are required) is postulated as key step of the formation of the minor lactone **168**.

The relationship between the reaction time and the enantiomeric purity of **167** (it decreased as the required reaction time becomes larger) was explained on the basis of a competition between two different mechanisms for the initial cycloaddition. The first involved the expected intermolecular Diels-Alder reaction, which must be completely stereoselective, exhibiting a similar course to that previously postulated for the reaction of **161A** and NMM [142] and depicted in Fig. 15. As this reaction is a bimolecular process, it must be strongly favored by high pressures. The competing mechanism would be the result of a three-step sequence, a slow intermolecular acylation of the sulfinyl group by maleic anhydride, followed by an intramolecular Diels-Alder reaction of the resulting acyloxy sulfonium intermediate, yielding compound **A′**, which is converted into **167′**, the enantiomer of **167** (Fig. 16).

The rather low reactivity of 1-sulfinyl dienes, requiring long reaction times or high pressures to react with good dienophiles, contrasts with the significant reactivity of 2-sulfinyl dienes, which are completely transformed into the adducts in a few hours under mild conditions. Taking into account the electron-withdrawing character of the sulfinyl group, its negative influence on the reactivity of dienes was not unexpected. Nevertheless, the fact that this effect was dependent on the relative position of the sulfinyl group is difficult to explain. In order to understand its unexpectedly low influence of the sulfinyl group on the dienophilicity of double bonds, we proposed that it could also act as an electron-

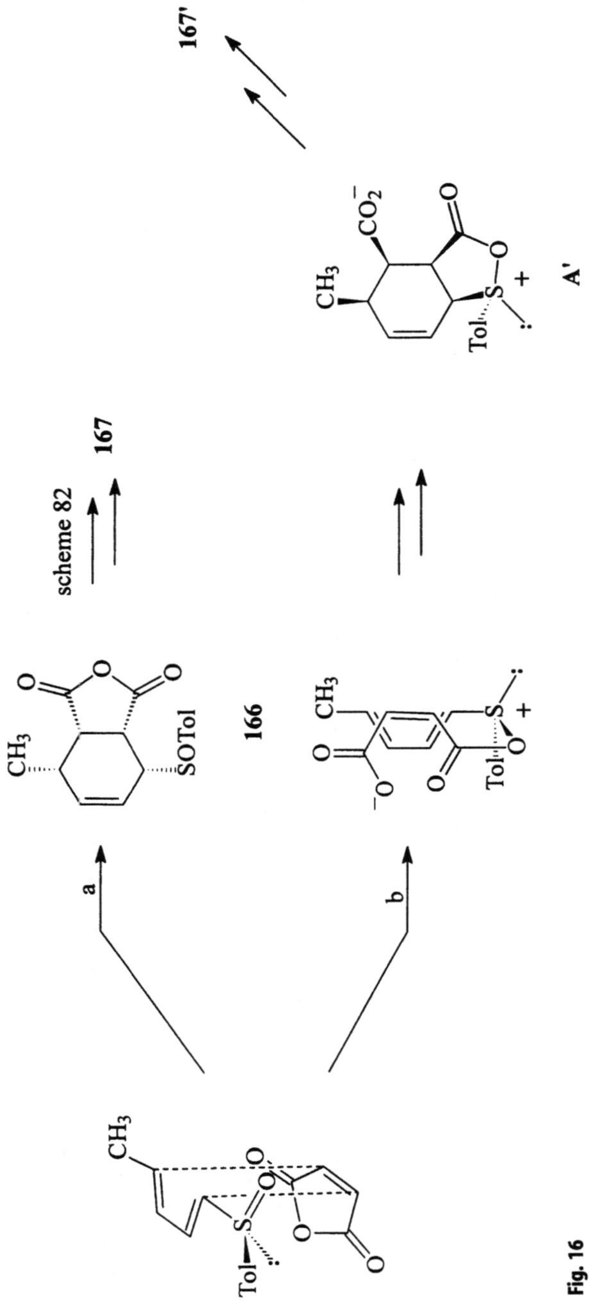

Fig. 16

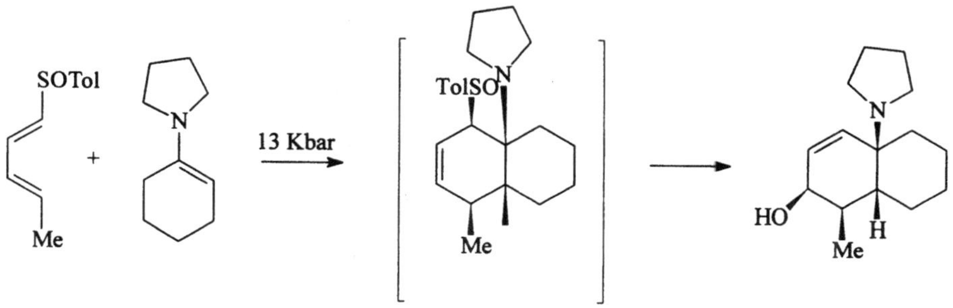

161a **169** **170** (*ee* = 28%)

Scheme 83

donating group, thus compensating its presumably activating effect. The same assumption could also explain that the reactivity of 2-sulfinyl dienes was not as low as could be expected by assuming that it merely exhibits an exclusive electron-withdrawing effect. The lower reactivity of 1-sulfinyl dienes could be related to the fact that they exhibited an extended conjugation instead of the crossed one of the 2-sulfinyl dienes, thus determining a higher efficiency of the sulfinyl group to withdraw electrons in 1-sulfinyl dienes.

The results obtained in reactions of methyl acrylate and enamines with diene **161a** are quite interesting from a mechanistic point of view [143b]. Starting from enamines derived from cyclohexanone and pyrrolidine, reactions require high pressure (13 kbar) and long reaction times (120 h), yielding cyclohexenol **170**, resulting from tandem Diels-Alder reaction/[2,3]-sigmatropic rearrangement (Scheme 83). The exclusive formation of **170** reveals that both regioselectivity and *endo* selectivity of the cycloaddition are complete, affording the adduct **169**, which undergoes a further rearrangement. The low optical purity of **170** (*ee* = 28%) suggested a poor π-facial selectivity.

The reaction of **161a** with ethyl acrylate also required the use of high pressures (13 kbar), and a mixture of six different adducts, with little synthetic value, could be isolated (Scheme 84). The regioselectivity of the reaction was moderate (4:1), the adducts **171** being favored (the orientating character of the methyl group is higher than that of the sulfinyl one). The proportion of the adducts indicates a moderate *endo*-selectivity (*endo*/*exo* = 55:22) and a very low π-facial selectivity for both *endo* (29:26) and *exo* (12:10) approaches yielding **171** (Scheme 84). By contrast, a complete *endo* selectivity and a higher π-facial selectivity (19:4) were observed for cycloadditions yielding regioisomers **172**.

Two main conclusions can be drawn from these results. First, the sulfinyl group has an unusual effect on the reactivity of the diene, which has a negative influence on reactions with both electron rich and electron poor dienophiles. Second, π-facial selectivity is efficiently controlled only in those cases where the regioselectivity of the reactions allows an interaction of the sulfinyl group at C(1) with substituents at dienophile-bearing electronegative heteroatoms in the TS affording *endo* adducts. In order to justify the low reactivity of 1-sulfinyl dienes, simultaneous increase of the LUMO's energy and decrease of the HOMO's energy at the diene, produced by +M and –I effects of the sulfinyl group, were proposed. This would explain the low reactivity of sulfinyl dienes in both Diels-Alder and inverse demand Diels-Alder reactions. The influence of the regioselectivity on the π-facial selectivity was rationalized by assuming that electrostatic repulsion between sulfinyl oxygen and the appropriate dienophilic substituents (the carbonyl oxygen of maleimides, maleic anhydride, and acrylates),

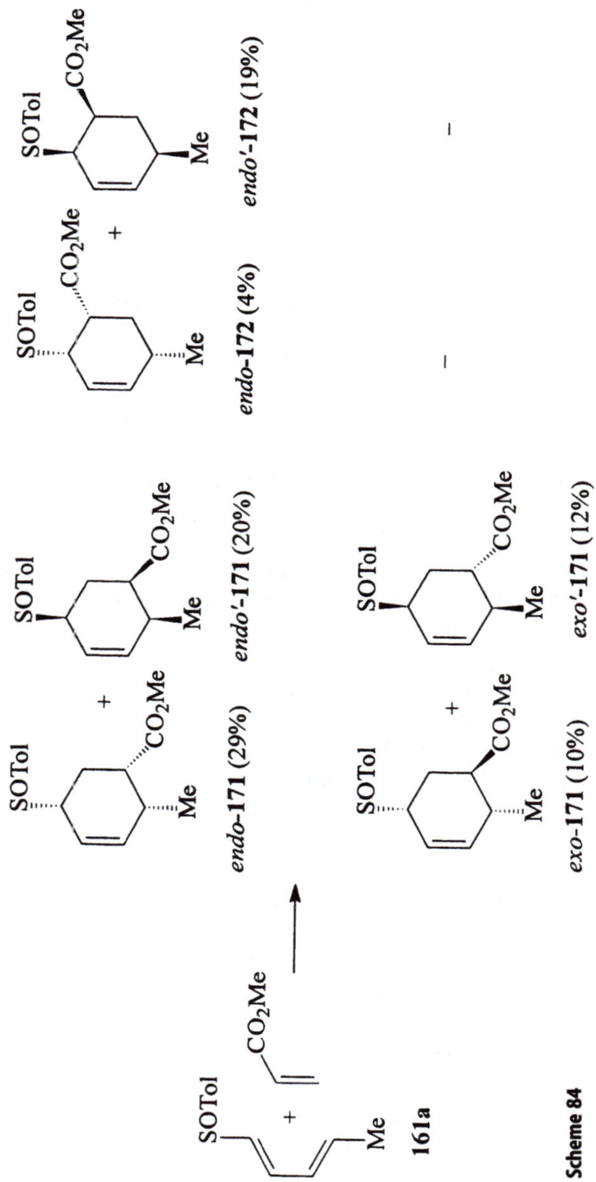

Scheme 84

restricts the conformational equilibrium around the C-S bond. This induces a shifting toward the rotamer exhibiting the longest distance between both oxygens in the transition state (s-*trans* arrangement of the sulfinyl oxygen). The favored approach of the dienophile must take place from the less hindered face of the diene (that bearing the lone electron pair) adopting such a conformation (**A** in Fig. 15). When there is no restriction to the conformational preferences, as happens in reactions with enamines or acrylates (in the TS affording compounds **171**), the π-facial selectivity is significantly decreased.

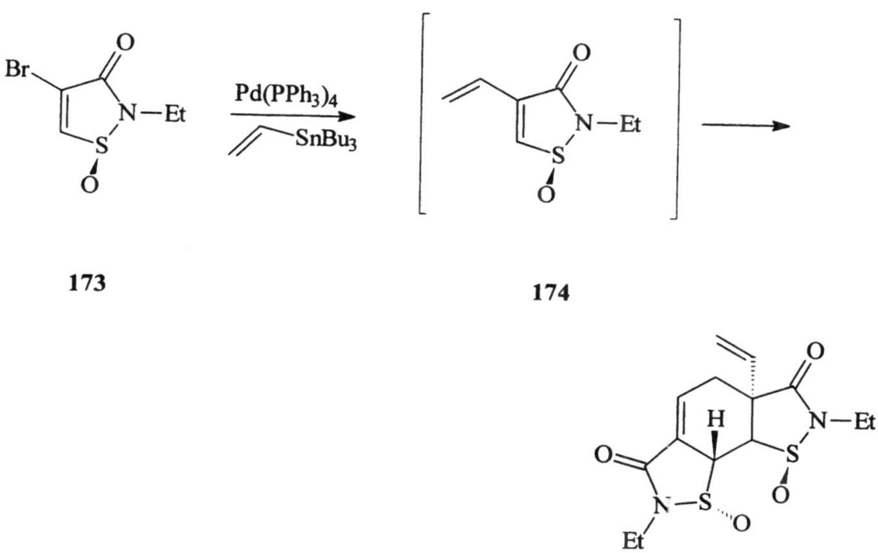

Scheme 85 **175**

Finally, dimerization of compound **174**, which is not really a sulfoxide but exhibits a structure similar to that of 1-sulfinyl dienes, has also been reported [144]. Coupling of **173** with vinyltributyltin yielded dimerized compound **175** instead of its precursor adduct **174** (Scheme 85). The stereochemistry of **175** suggested that it could result from an *exo* approach of the dienophile to the most hindered face of the diene (the opposite one to that bearing the lone pair) which was explained by assuming that chelation may play an important role in the stereoselectivity. The high reactivity exhibited by **174**, acting simultaneously as diene and dienophile, is noteworthy, and could be related to the fact that SO group is contained in a cyclic structure, which makes the conjugation of the lone electron pair at sulfur with the double bond difficult (+M effect of the SO group), thus increasing both dienophilic and dienic reactivity of **174**.

3 Hetero Diels-Alder Reactions

Sulfinyl dienes and vinyl sulfoxides have rarely been used in asymmetric hetero-Diels-Alder reactions [145]. The first example was reported in 1992 and describes an intramolecular cycloaddition using a heterodiene bearing a chiral sulfinyl group [146a]. In this paper, the conversion of *α*-*p*-tolylsulfinyl *α*,*β*-unsaturated ketone **176** (prepared by Knoevenagel reaction of 3-methylcitronellal and (*S*)-*p*-toluenesulfinylacetone) into the hetero-Diels-Alder adducts **177**

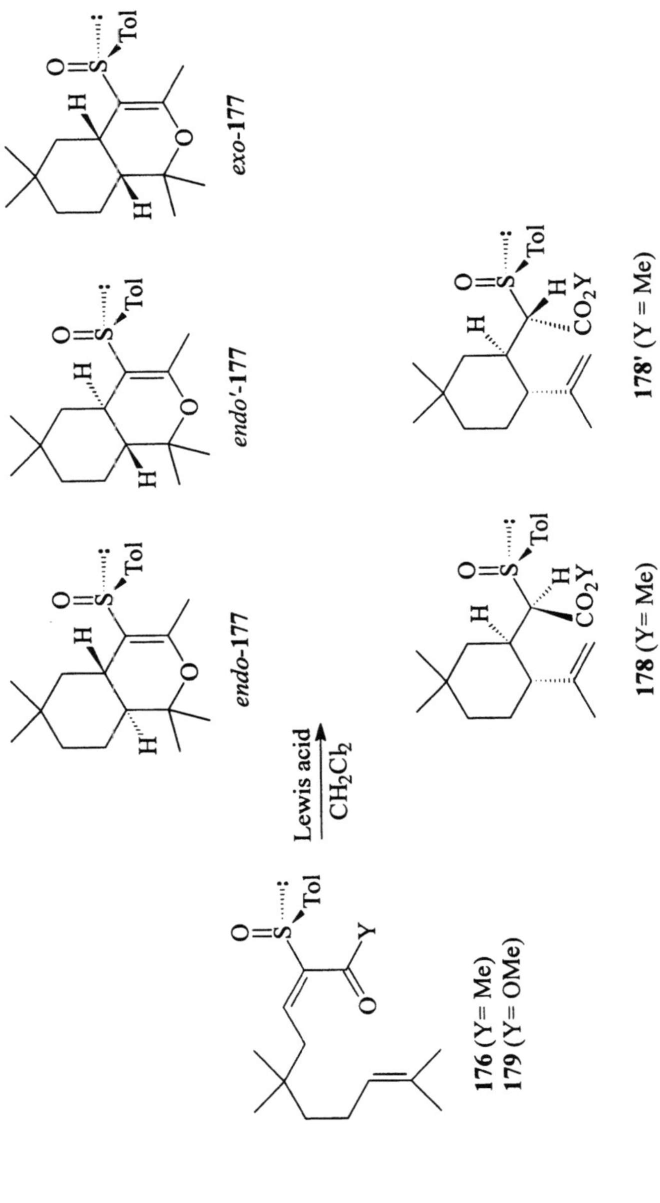

Scheme 86

under Lewis acid catalysis in CH_2Cl_2 at low temperature (–78 °C to around 0 °C) is reported (Scheme 86). In some cases, the products of the *ene*-reaction **178** were also formed. Monodentate Lewis acids, such as Et_2AlCl, $EtAlCl_2$, and $BF_3 \cdot OEt_2$, afforded only **177** in high yields (> 90 %). The π-facial selectivity was very high for the *endo* approach (*de* ~ 80 %), and the *endo*/*exo* ratio became 96:4 with Et_2AlCl. The use of bidentate Lewis acids, such us ZnX_2 or $SnCl_4$, provided significant amounts of **178**, which became the major products with ZnX_2 catalysis. The asymmetric induction observed in ene reactions catalyzed by $SnCl_4$ was almost complete (**178**:**178′** = 95:5). The use of compound **179** as the starting material gave a mixture of only **180** and **180′** with both monodentate and bidentate catalysts, diastereoisomeric excesses being slightly lower [146b].

On the basis of the stereochemical results obtained the authors proposed that the most stable conformation of heterodiene in the TS was that exhibiting the bulkiest tolyl group oriented *anti* to the acetyl group (see Fig. 17). Dienophile approach takes place to the less hindered face of heterodiene containing the lone electron pair at sulfur, yielding *endo*-**177** as the major adduct. The chelation of the carbonyl and sulfinyl oxygens by bidentate Lewis acids precluded the s-*cis* arrangement of the C = C and C = O bonds required for the hetero-Diels-Alder reaction. This was invoked to explain the increase in the proportion of the products **178** resulting from the *ene* reactions.

Fig. 17

The first example of an intermolecular hetero-Diels-Alder reaction of a heterodiene bearing a chiral sulfinyl group was due to Maignan et al. [147]. Treatment of (+)-(*S*)-3-*p*-tolylsulfinyl-3-buten-2-one **181** with 2-methylentetrahydrofuran gave a 1:1 mixture of the spiroketals **182a** and **182b** (Scheme 87). The reaction was completely regioselective under mild conditions (room temperature, 3 h), but the π-facial diastereoselectivity was very low.

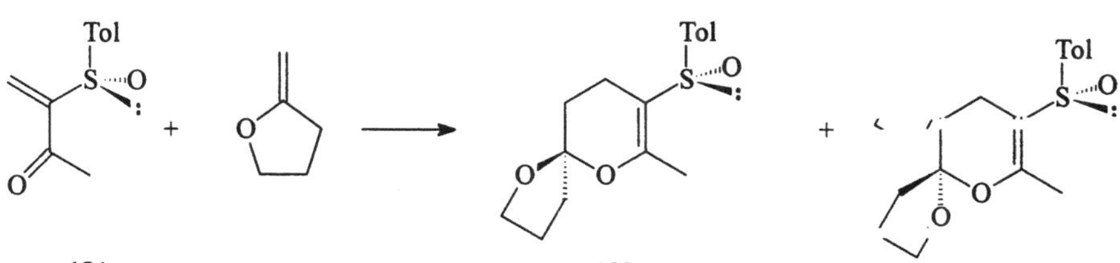

181 **182a** **182b**

Scheme 87

Similar results were observed in reactions with acyclic vinyl ethers (*de* ranged between 0% and 14%) affording **183**, whereas reactions of **181** with styrenes were surprisingly highly diastereoselective, yielding a mixture of adducts **184a** and **184b** (Scheme 88) with high optical yield (*de* > 90%) [148]. The authors do not explain the reasons for this unexpected and interesting behavior exhibited by the styrenes.

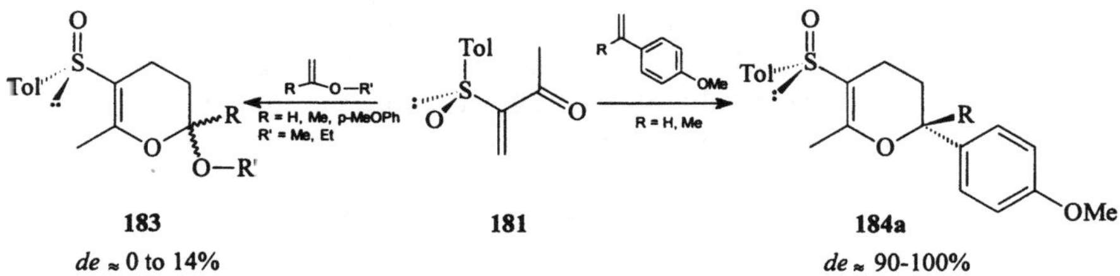

Scheme 88

The hetero-Diels-Alder reaction between sulfinyl diene **152a** and ethyl glyoxalate [149] has also been reported. (Scheme 89). In the presence of $LiClO_4$, the *endo* adducts **185** are predominant, but under $Eu(fod)_3$ or $ZnCl_2$ catalysis the *exo* adducts **186** are the major ones. The *endo/exo* selectivity is rather moderate, and the π-facial selectivity is low in all cases.

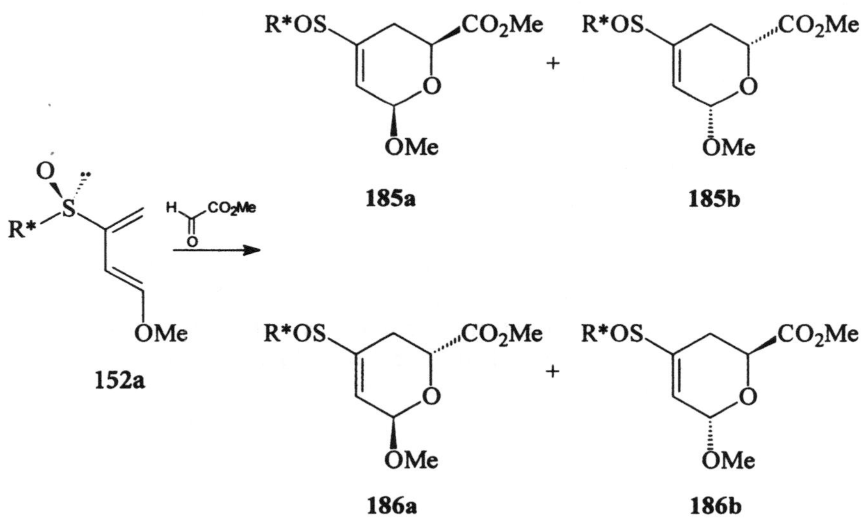

catal.	Yield (%)	185a	185b	186a	186b
-	80	45	28	18	9
$Eu(fod)_3$	52	59	21	14	6
$LiClO_4$	82	9	5	61	25
$ZnCl_2$	40	9	9	39	43

Scheme 89

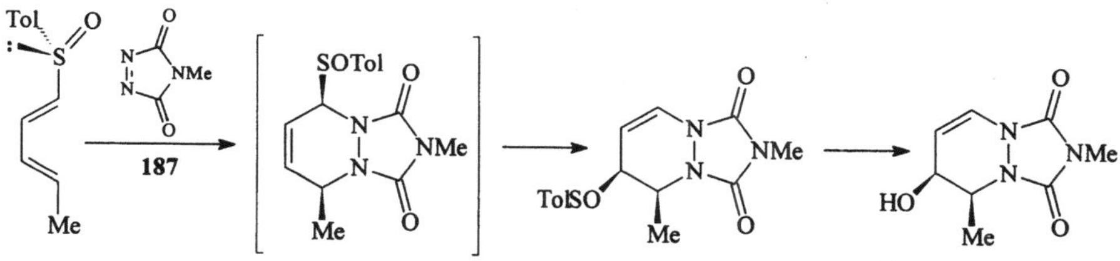

161a *endo*-**188** **189** **190**

Scheme 90

We have recently found that hetero-Diels-Alder reaction between (*R*)-1-(*p*-methylsulfinyl)-1,3-pentadiene (**161a**) and 4-methyl-1,2,4-triazoline-3,5-dione (**187**) under very mild conditions (–40 °C to –10 °C, 2 h) afforded sulfenate **189**, which is readily transformed into the carbinol **190** by treatment with $P(OMe)_3$ as a thiophilic agent [150]. Compound **189** is presumably derived from cycloadduct **188** (which is not detected) through a stereocontrolled sigmatropic [2,3]-rearrangement. The optical purity of **190** (ee > 99 %) confirmed that both the initial cycloaddition and the subsequent rearrangement took place with complete control of the π-facial selectivity. The absolute configuration of **190** was determined to be that depicted in Scheme 90 from NMR studies of its (*R*)- and (*S*)-MTPA esters. This in turn allowed the assignment of the absolute configuration of the cycloadduct *endo*-**188**, which suggested a stereochemical course identical to that proposed for the reaction of **161a** with *N*-methyl maleimide (see 1-sulfinyl dienes). The high reactivity of compound **187** as heterodienophile will allow us to take advantage of the excellent properties of 1-sulfinyl dienes as chiral dienes, which have so far been severely limited by its low reactivity with the usual dienophiles.

4
1,3-Dipolar Cycloadditions

1,3-Dipolar cycloadditions provide a useful method for preparation of a wide range of five-membered ring heterocycles [151]. The potentially high degree of stereocontrol, a consequence of a concerted mechanism [152], reinforces their importance in asymmetric synthesis [151e]. Vinyl sulfoxides have been widely used as synthetic equivalents of acetylenes in 1,3-dipolar cycloadditions in order to obtain a range of aromatic heterocycles. Nevertheless, their use as enantiomerically pure dipolarophiles in asymmetric processes has been much less studied than their use in Diels-Alder reactions; in many cases this may be due to the easy desulfinylation of the resulting cycloadducts to afford aromatic compounds. In the sections below we will consider these reactions, grouped according to the nature of the dipoles.

4.1 Nitrones

Nitrones were the first as well as the most widely used dipoles in asymmetric cycloadditions. The first report on the use of enantiomerically pure vinylsulfoxides as dipolarophiles was due to Koizumi et al. [153], who described in 1982 the reaction of (*R*)-vinyl *p*-tolyl sulfoxide **1** with acyclic nitrones **191**. The reactions required 20 h in refluxing benzene to be completed, yielding a mixture of only two compounds, **192** and **193** (Scheme 91). They exhibited identical *endo* or *exo* stereochemistry (which was not unequivocally assigned), deduced from the fact that their reduction yielded enantiomeric thioethers. The major component, **192**, exhibits (*S*) configuration at C-3, determined by chemical correlation. The authors claim this paper [153] to be the first example of 1,3-dipolar cycloaddition using chiral dipolarophiles.

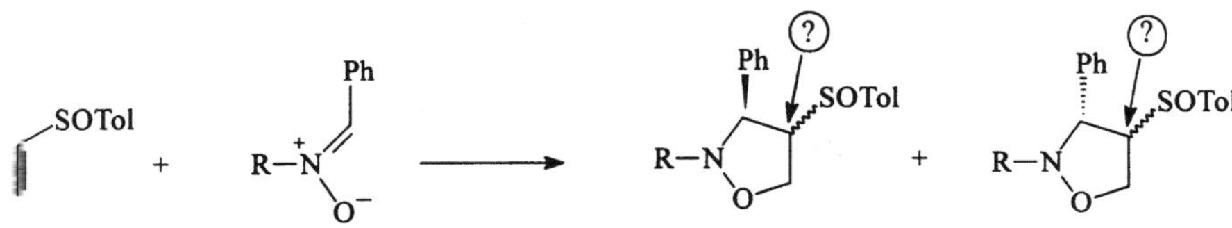

1 **191a** (R=Me) **192b** (R=Ph) **192** (major) **193** (minor)

Scheme 91

These results indicate that the sulfinyl group seems to be much more efficient in the control of the stereoselectivity of 1,3-dipolar cycloadditions (*endo* or *exo* adducts are exclusively obtained in *de* > 80%) than in Diels-Alder processes (mixtures of all four possible adducts were formed). Additionally, complete control of the regioselectivity of the reaction was observed. Despite these clearly excellent results, the following paper concerning asymmetric cycloaddition of cyclic nitrones and optically pure vinyl sulfoxides was reported nine years later [154]. (Meanwhile, only one paper [155], related to the synthesis of *β*-nicotyrines, described the use of reaction of nitrones with racemic vinyl sulfoxides, but these substrates were merely used as a masked equivalent of acetylene dipolarophile). In 1991, Koizumi et al. described the reaction of one of the best dipolarophiles, the sulfinyl maleimide **109**, with 3,4,5,6-tetrahydropyridine 1-oxide **194** [154]. It proceeded in CH_2Cl_2 at –78 °C to afford a 60 : 20 : 10 : 6 mixture of four products in ca. 90 % yield (Scheme 92).

The major adduct *exo*(*c*)-**195** was isolated by crystallization from the reaction mixture (33% yield) and its stereochemistry was unequivocally established as *exo* with respect to the sulfinyl group by X-ray diffraction studies. The other three adducts could not be separated, but traces of one of them could be identified as a regioisomer of compound **195**. From this fact and consideration made on the ^{1}H NMR spectra obtained from the reaction mixture, the authors conclude that this reaction takes place with a high diastereoisomeric excess, the *exo* adducts being favored. This stereochemical course was explained by assuming that the approach of the 1,3-dipole takes place from the face supporting the lone

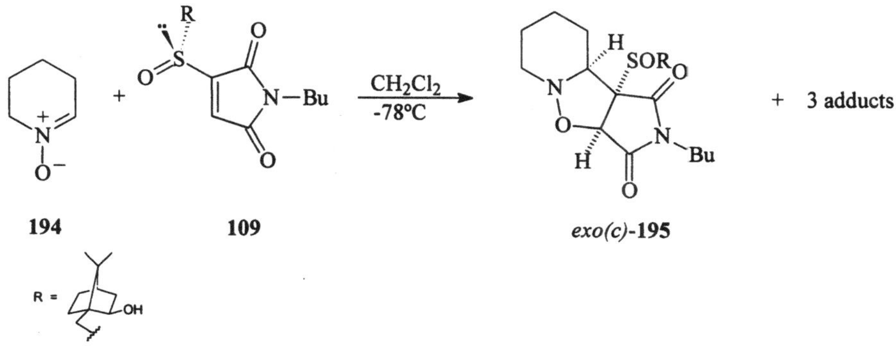

Scheme 92

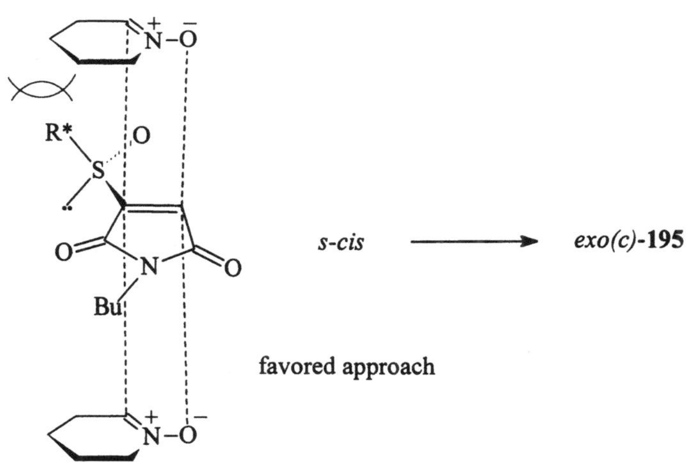

Fig. 18

electron pair at sulfur in the *s-cis* conformation of the dipolarophile, i.e., the same proposal used to rationalise the behavior of **109** as a dienophile (Fig. 18).

Reactions of cyclic nitrone **194** with homochiral alkynyl sulfoxides **8** have also been studied [156]. They were conducted at room temperature (24 h), affording 1:1 mixtures of the diastereoisomeric isoxazolines **196a** and **196b** in excellent yields (Scheme 93). These compounds, which derived from the approach of dipolarophiles to each face of the 1,3-dipole, exhibit large differences in their R_f values, which allowed their ready separation by flash chromatography as a previous step to their independent hydrogenation into enantiomerically pure piperidines **197a** and **197b**. The dramatic effect of the sulfinyl group, increasing both reactivity and regioselectivity of alkynes with nitrones, is remarked in this paper. The regioselectivity was as expected according to the electron-withdrawing character of the sulfinyl group.

The asymmetric 1,3-dipolar cycloaddition of *N*-methyl-3-oxo pyridinium **198** (R = H) with (*R*)-*p*-tolyl vinyl sulfoxide (**1**) has been studied [154, 157a]. The

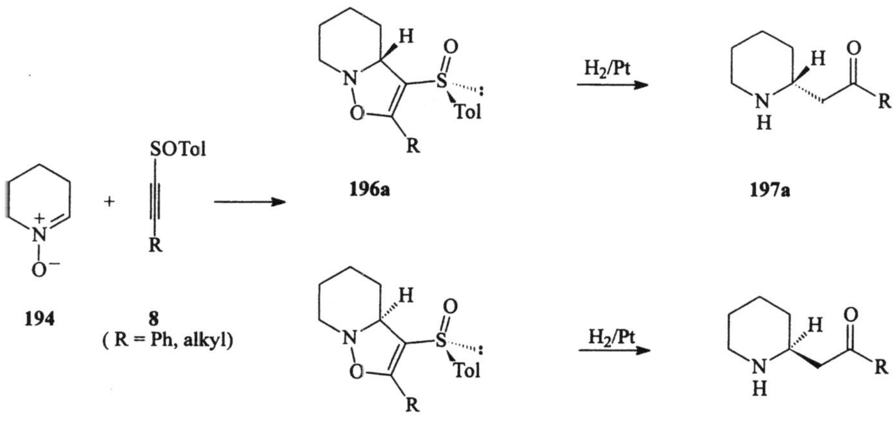

Scheme 93

reaction is moderately *exo* selective (*exo*:*endo* = 60:40) and the π-facial selectivity is high for the *exo* approach (*de* = 68%) and complete for the *endo* one (Scheme 94). The major adducts are those resulting from attack of the dipole at the less hindered face of the dipolarophile in the *s-trans* conformation. Compound *exo*(*t*)-**199** was used to prepare enantiomerically pure (1*S*)-(–)-2α-tropanol. Similar stereochemical behavior has been found for **198′** (R=Ph) [157b] as indicated in Scheme 94. The obtained adducts were transformed into 2-alkyl-3-phenyl substituted tropanes.

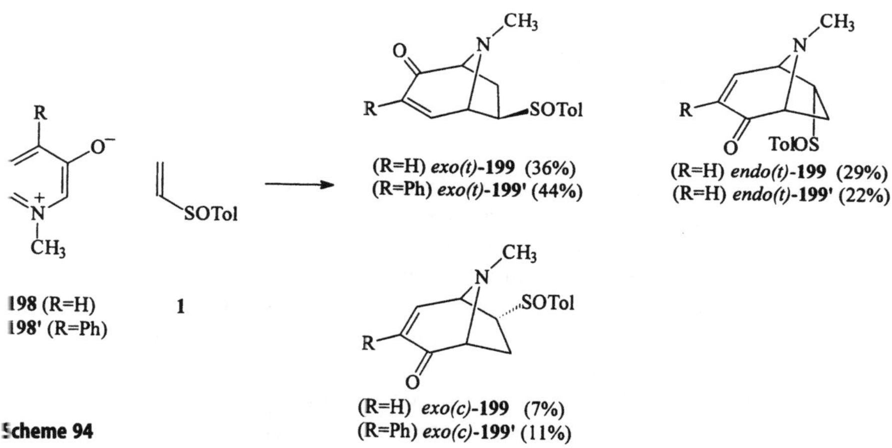

Scheme 94

More recently, Bravo et al. [158] described the synthesis of optically pure fluoro-substituted isoxazolidines by 1,3-dipolar cycloaddition of acyclic nitrones to chiral methyl enol ethers of 3-fluoro-1-sulfinyl-2-propanones. Reaction of **191b** with **200** (room temperature, 10 days) afforded only one adduct, *exo*(*t*)-

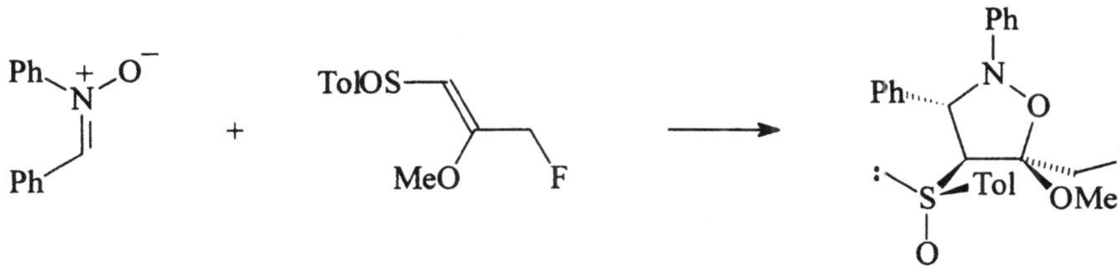

191b **200** *exo(t)*-**201**

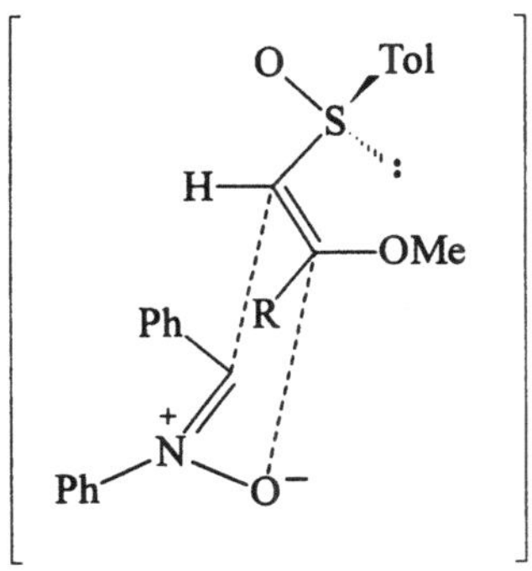

Scheme 95

201, in 90% yield (Scheme 95). This result indicates a total control of the regioselectivity, *exo* selectivity, and even π-facial selectivity. Other nitrones (*N*-methyl, *C*-phenyl and *N*-benzyl, *C*-ethoxycarbonyl) as well as other enol ethers (enolates do not react with nitrones) gave similar results.

The stereochemical course of these reactions was also explained by assuming an *exo* approach of dipole to the vinyl sulfoxide in *s-trans* conformation, which in this case would be the less destabilized by electrostatic repulsions (Scheme 95). A similar stereochemical course would explain the results obtained in the reactions of nitrone **194** with (*Z*)-vinyl sulfoxides **13** (Scheme 96) [159a]. With these dipolarophiles, the *exo* selectivity is complete, and the π-facial selectivity is very high (*de* ≈ 82–98%) and depends on the size of the R group, which must be responsible for the shifting the conformational equilibrium around the C-S bond toward the *s-trans* rotamer. The major adduct *exo*(*t*)-**202** (R = Me) was transformed into the enantiomerically pure piperidine alkaloid (+)-sedridine.

The π-facial selectivity of cycloadditions between (*Z*)-**13** and the five-membered cyclic nitrone **194′** was slightly lower (*de* ≈ 64%). The pyrrolidine natural compound (–)-hygroline was obtained from the major adduct, *exo*(*t*)-**202′** [159b]. Both *exo* and π-facial selectivities decreased in reactions with vinyl sulfoxides of (*E*)-configuration. The stereochemical results obtained from (*Z*)-**13** were explained by assuming the *exo* approach of the dipole to the less hindered face of the dipolarophile **13** which adopts conformation **A** depicted in Scheme 96, with the lone electron pair at sulfur in an *s-cis* arrangement. This explana-

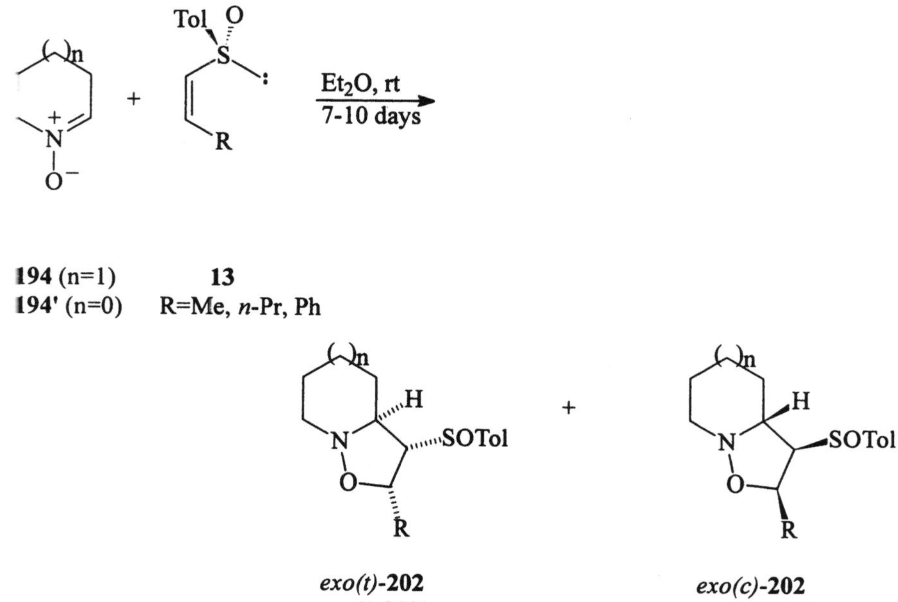

Scheme 96

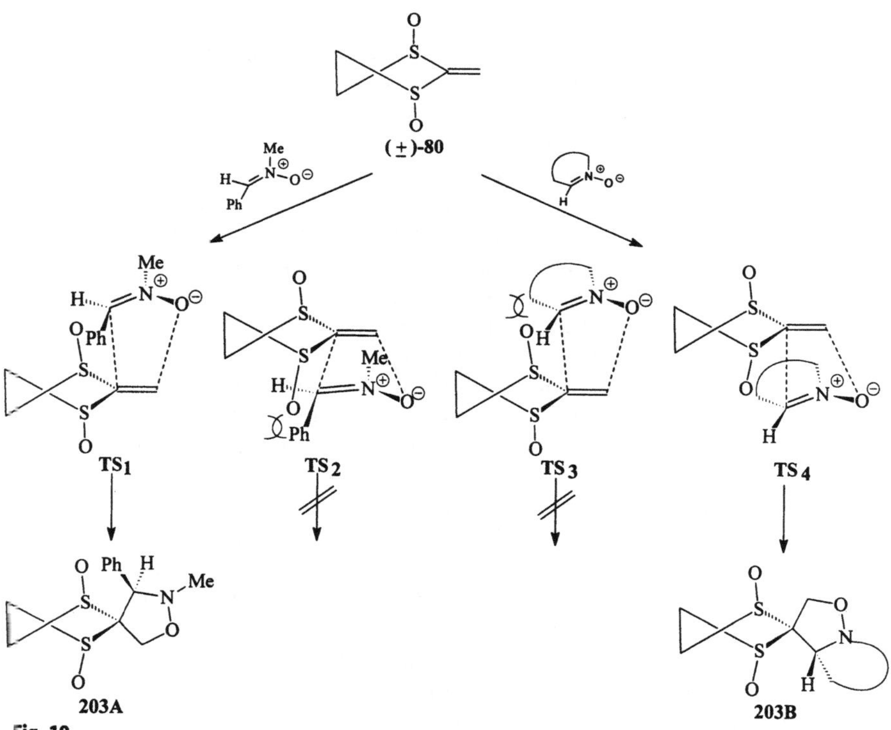

Fig. 19

tion was based on the conformational preferences of vinyl sulfoxides obtained by ab initio calculations (6–31G*) [160].

As a conclusion to the results so far reported, vinyl sulfoxides show good reactivity toward nitrones, and the stability of the resulting sulfinyl cycloadducts allows their easy isolation. Both the regioselectivities and *exo*-selectivities of these reactions are usually very high, as well as their π-facial selectivity, which in the case of the acyclic sulfoxides is seemingly related to their ability to adopt the *s-trans* conformation.

In a recent paper [161], Aggarwal et al. have reported that racemic *trans*-2-methylene-1,3-dithiolane, (±)-**80** (see Scheme 43) adds to both cyclic and acyclic nitrones with very high stereoselectivity and complete regioselectivity to give 4,4-disubstituted isoxazolidines **203A** and **203B**. The two possible transition states are depicted in Fig. 19. The higher stability of **TS1** and **TS4** with respect, to **TS2** and **TS3**, was explained upon the basis of steric interactions, thus rationalizing the formation of the major adducts, indeed exclusive in many cases. The unexpectedly complete regioselectivity induced by this dipolarophile, yielding 4,4-rather than 5,5-disubstituted isoxazolidines, is also described in the paper.

4.2 Nitrile Oxides

The first report of reactions of this kind of dipoles with racemic vinyl sulfoxides was published in 1973 [162]. Benzonitrile oxide reacts smoothly with sodium salts of some β-ketosulfoxides, **204**, to give the corresponding 4-methylsulfinyl isoxazoles **205** in good yields. The isoxazoline intermediates were not isolated in any case. To our knowledge, this was the first paper reporting the use of vinyl sulfoxides as dipolarophiles and revealed the strong tendency of the resulting isoxazoline to be transformed into aromatic isoxazoles. The reaction of nitrilimine **206** with sulfinyl enolates **204** to obtain sulfinyl pyrazoles **207** was also reported in this paper (Scheme 97).

Further studies on racemic sulfoxides demonstrated that the sulfinyl group had only a moderate influence on the regioselectivity. Thus, an investigation of cycloadditions of nitrile oxides to methyl styryl sulfides, sulfoxides, and sulfones [163] revealed that sulfenyl and sulfonyl derivatives reacted with almost complete (but opposite) regioselectivity, wheras sulfoxides yielded mixtures of regioisomeric isoxazolines, which were easily transformed into aromatic isoxazoles. This behavior was rationalized by theoretical calculations as a result of the influence of the thio moieties on the FMO. Moreover, the stereoselectivity of reactions of mesitonitrile oxide with racemic benzothiophene *S*-oxides **208a** (R = Me [164]) and **208b** (R = Ph [165]) was found to be very poor, yielding 1:1 mixtures of the two possible adducts (*syn*-**209** and *anti*-**209**) resulting from the approach of the dipole to either diastereotopic face of **208** (Scheme 98). We must point out that reactivity of **208b** toward these dipoles was found to be similar to that of the corresponding sulfone [165], which contrasts with the results obtained in Diels-Alder reactions, where vinyl sulfones clearly exhibited a higher reactivity than that of vinyl sulfoxides.

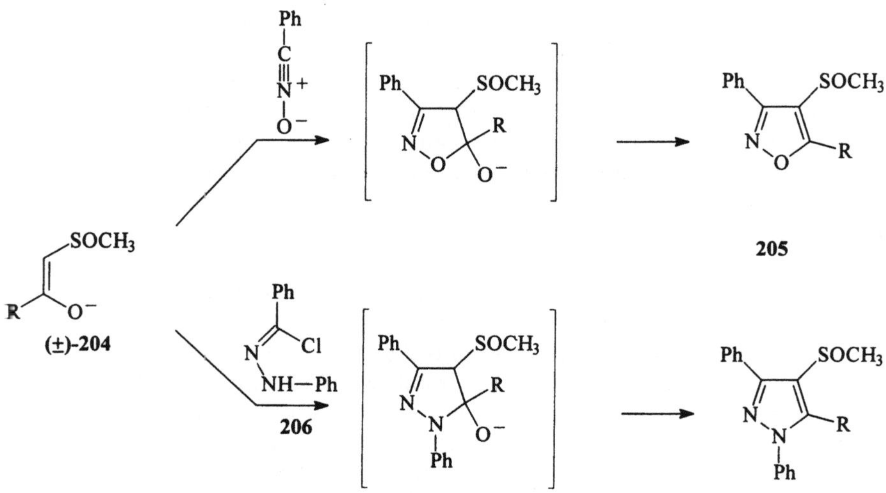

Scheme 97

Given the results described above, it is not surprising that use of these reactions in asymmetric synthesis is rare. In 1993, Bravo et al. used this methodology to synthesize interesting fluorosubstituted heterocycles (**211**, R = CFR^{1}R^2) starting from enantiomerically pure sulfoxides [166a]. In cases where the dehydration process was much slower, and isoxazoline intermediates could be isolated (R = CFMePh, Scheme 99), a single diastereoisomeric 5-hydroxy-4,5-dihydroisoxazole **210** was observed, suggesting a complete π-facial selectivity. Nevertheless, the ready transformation of these cycloadducts into the corresponding aromatic heterocycles **211**, lacking asymmetric carbons, decreased the synthetic interest of these reactions from the perspective of asymmetric synthesis.

The results obtained using enol ethers derived from enantiomerically pure β-ketosulfoxides as substrates were more interesting due to the higher stability of

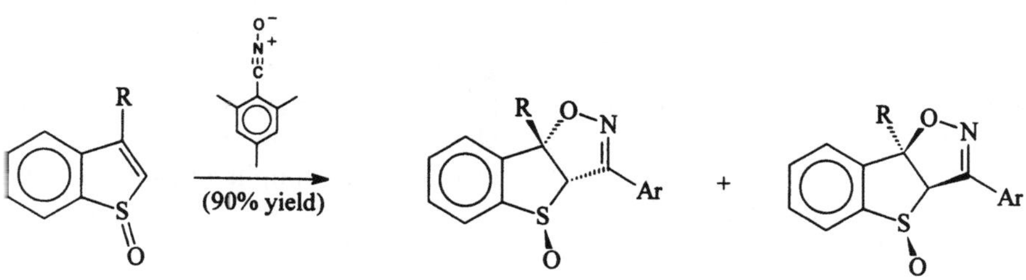

208a (R = Me)
208b (R = Ph)

syn-**209**

anti-**209**

Scheme 98

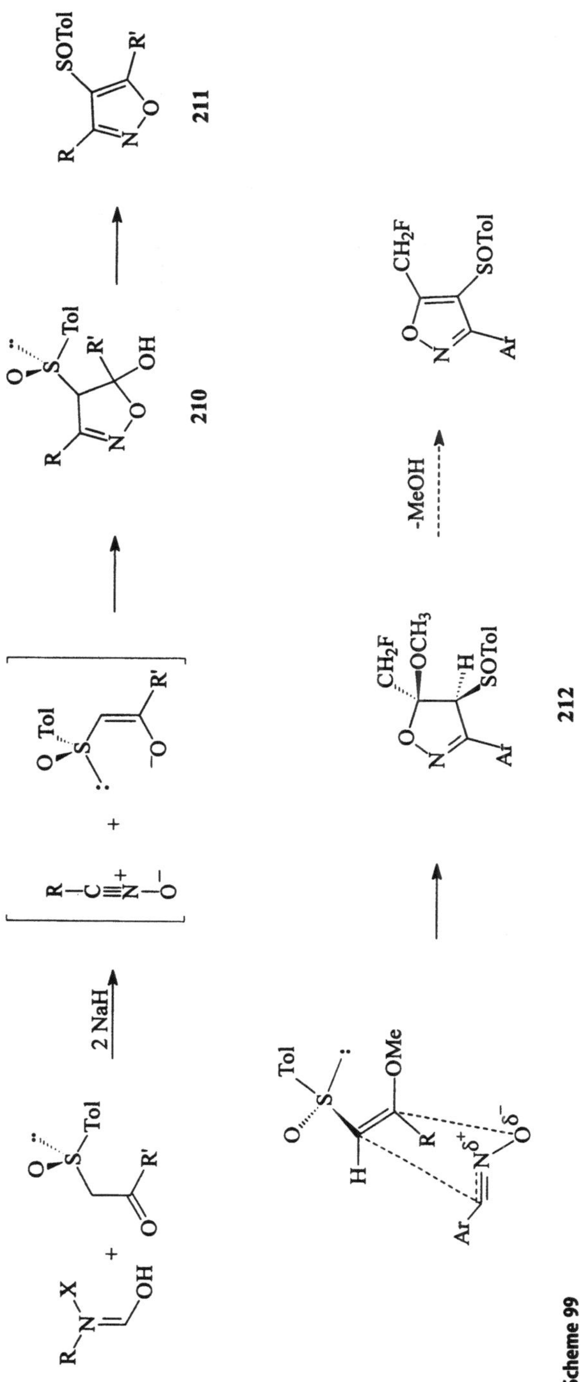

Scheme 99

the resulting cycloadducts **212** (Scheme 99) [166b]. Their isolation and characterization allowed confirmation of the complete π-facial selectivity of these reactions. The stereochemistry of the isolated compound suggests that it could be the result of the approach of the dipole toward the less hindered face of the dipolarophile in the *s-trans* conformation (the most stable due to electrostatic interactions).

The last results reported in this field are related to the 1,3-dipolar reactions of *tert*-butyl (*E*)-4,4-diethoxy-2-*p*-tolylsulfinylbut-2-enoate and (S_5,S_S)- and (R_5,S_S) 5-ethoxy-3-*p*-tolylsulfinylfuranones (**42a** and **42b**) with different nitrile oxides [167]. Acyclic sulfoxides react with benzonitrile, acetonitrile, and bromoformonitrile oxides to yield isoxazoles resulting in desulfinylation from the adducts. Cyclic dipolarophiles afford bicyclic isoxazolines in their reactions with benzonitrile oxide. The reactivity of the double bond as a dipolarophile is strongly increased by the sulfinyl group. The regioselectivity of these reactions

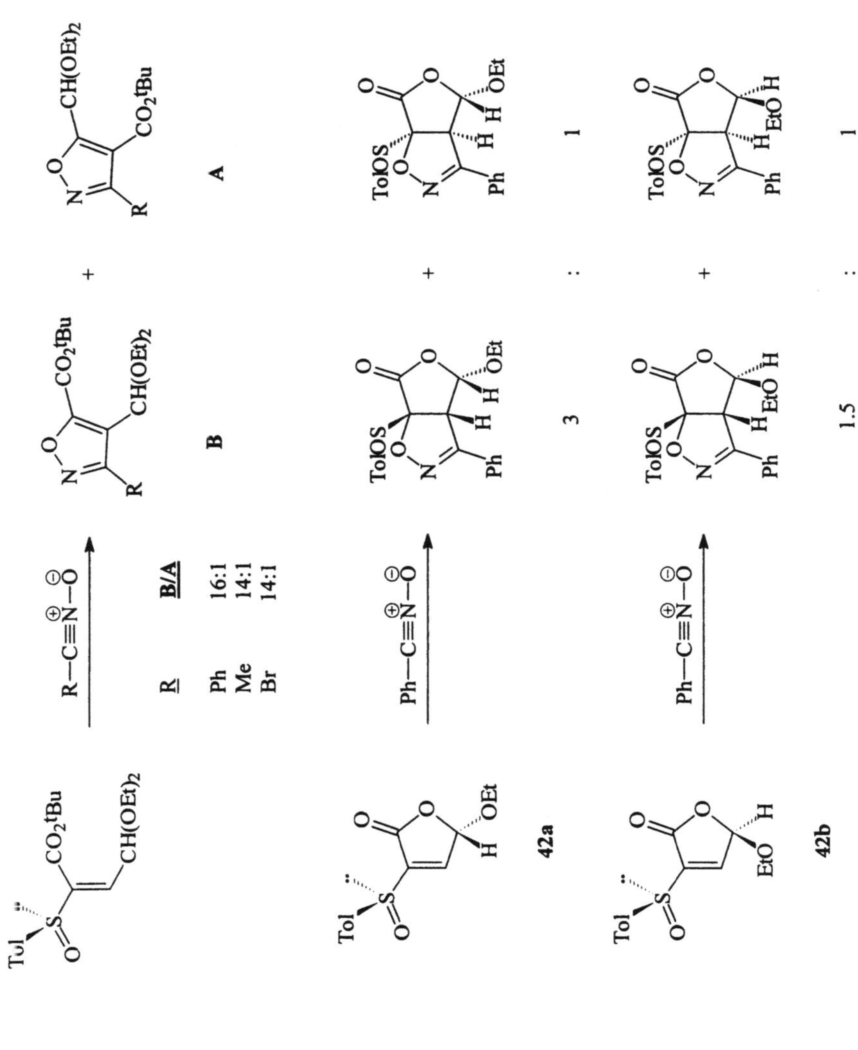

Scheme 100

is very high or complete and usually the opposite to that exhibited by dipolarophiles lacking the sulfinyl group. Electrostatic interactions and steric effects are invoked to explain this behavior. The π-facial selectivity of the reactions of **42a** (50% *de*) is higher than that of **42b** (20% *de*, see Scheme 100). As steric grounds would predict a more stereoselective evolution for **42b** (it was the case in Diels-Alder reaction, see Scheme 21), electrostatic attraction between the positively charged carbon atom at the dipole and the oxygen of the ethoxy group at the dipolarophile has been invoked to justify the higher diastereoisomeric excess observed in reaction from **42a**.

4.3 Diazoalkanes

Vinyl sulfoxides have been used as synthetic equivalents of alkynes in reactions with diazoalkanes to prepare pyrazoles. The initially obtained adducts subsequently eliminate or rearrange the sulfoxide moiety to achieve pyrazoles lacking the sulfur function. Thus, the adducts resulting by reaction of CH_2N_2 with the sulfinylated double bond of allenyl sulfoxides **213** are transformed through a sulfoxide-sulfenate rearrangement into hydroxymethyl pyrazoles **214** [168], whereas those obtained by reaction with sulfinyl coumarins **215** suffered sulfinyl elimination into the pyrazoles **216** [169]. In both cases 1,*H*-pyrazoles were obtained as a consequence of a final tautomerization step (Scheme 101). These studies were carried out on racemic sulfoxides.

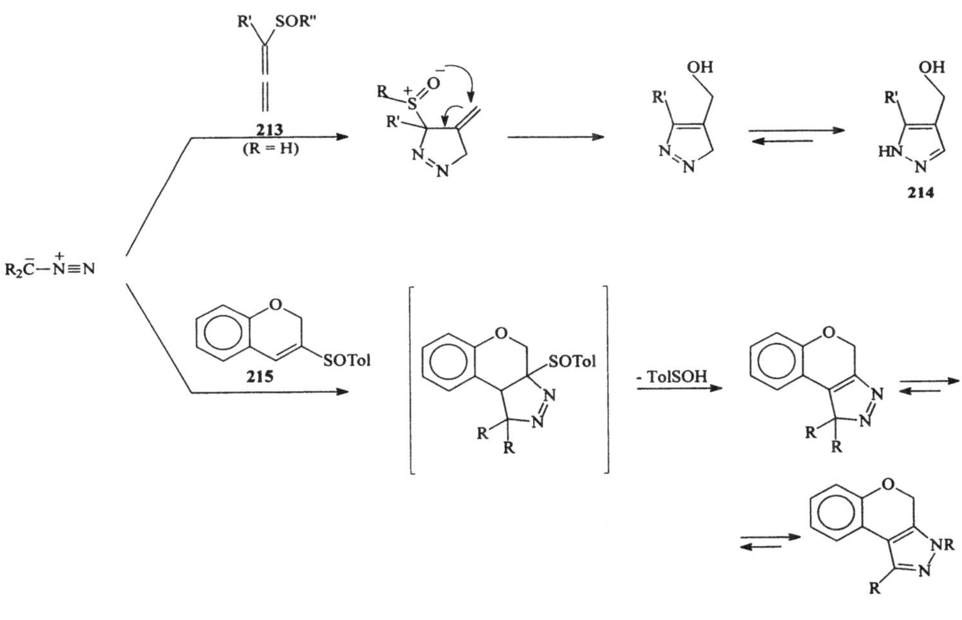

Scheme 101 **216**

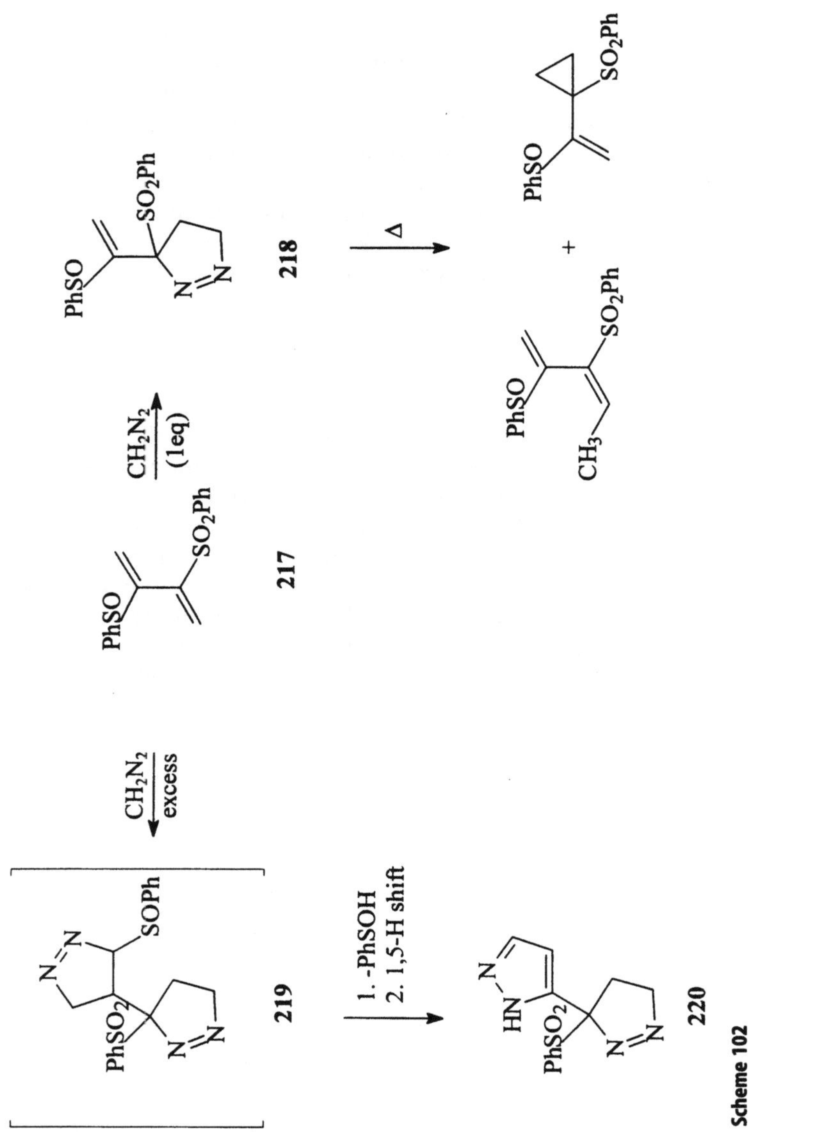

Scheme 102

The regioselectivity of these cycloadditions is completely controlled by the sulfinyl group. Unfortunately, sulfinyl elimination from the intermediate pyrazolines is faster than other much more interesting processes, such as extrusion of the nitrogen affording cyclopropanes or dienes. This has been nicely demonstrated by Padwa et al. in the reaction of 2-phenylsulfinyl-3-phenylsulfonyl-1,3-butadiene **217** with diazomethane [170]. The 1:1 cycloadduct **218**, formed from the sulfonylated double bond (more reactive than the sulfinylated one) readily extruded nitrogen, producing the related 1,3-dienes and cyclopropanes. On the other hand, bis-adduct **219**, formed with an excess of CH_2N_2, rearranged under the reaction conditions by desulfinylation and further 1,5-H shift into the 1,*H*-pyrazol **220** (Scheme 102).

A study of the reactions of diazoalkanes and *N*-benzyl azomethine ylid with α-thioderivatives (phenylsulfenyl, phenylsulfinyl, and phenylsulfonyl) of 3,3,3-trifluoropropene has been recently published by Viehe et al. [171] (Scheme 103). The reactions of sulfoxide **221** with diazoacetate and diazomethane regiospecifically yielded pyrazolines **222**, which could be easily transformed into 1,*H*-pyrazoles **223** by desulfinylation and further 1,5-*H* shift (Et_2O and benzene reflux respectively were used to transform cycloadducts derived from diazoacetate and diazomethane). Treatment of **221** with diphenyldiazomethane gave a mixture of the rearranged pyrazole **224** and cyclopropane **225** (Scheme 103). Reaction of **221** with *N*-benzyl azomethine ylid afforded pyrrolidine **226** in good yield. Although the authors do not comment on the stereochemical course of these reactions, seemingly only one stereoisomer of compounds **222** and **226** was obtained, which suggests a highly stereoselective process.

The only report on asymmetric cycloadditions of diazoalkanes and enantiomerically pure vinyl sulfoxides was published in 1996 [172]. Reaction of diazomethane with (S_S)-5-ethoxy-3-*p*-tolylsulfinylfuran-2(5*H*)-ones **42a** and **42b**, epimers at C-5, required 5 min at 0 °C to afford enantiomerically pure sulfinyl pyrazolines **227a** and **227b** in quantitative yields. These results show that the sulfinyl group strongly increases the reactivity of the butenolide with diazomethane (reactions of 5-alkoxybutenolides with CH_2N_2 required at least 12 h to be completed), which contrasts with its small influence in the case of Diels-Alder reactions of butenolide with cyclopentadiene [50]. Stabilizing electronic factors involving sulfinyl group and nitrogen were invoked to explain the strong posi-

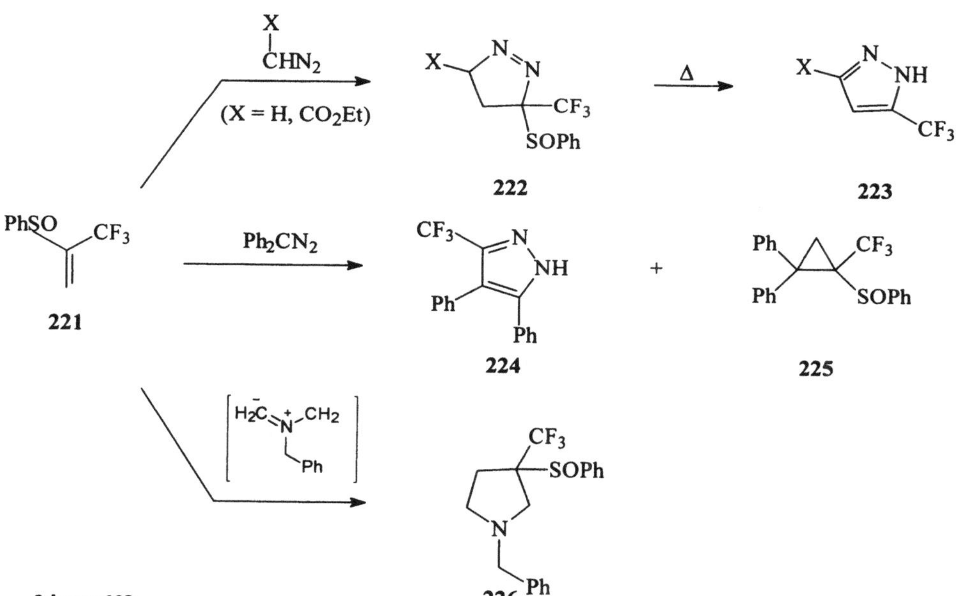

Scheme 103

tive influence of the sulfinyl group in the dipolarophilicity of the sulfinylfuran-2(5*H*)-ones. The regioselectivity of these reactions was complete, as expected from the behavior of other furanones, while their π-facial selectivity seemed to be exclusively controlled by the sulfinyl group. This conclusion was deduced from the fact that **227a** and **227b** are epimers at C-4, but they have identical configuration at C-3a and C-6a (unequivocally established by NMR spectroscopy). It was remarkable that the presence of the sulfinyl group in the 5-alkoxy butenolide allowed the complete control of the π-facial selectivity of these 1,3-dipolar reactions, which contrasts with the poor results obtained in reactions of 5-alkoxy furanones with cyclopentadiene (*de* < 20%) [50]. Moreover, the influence of the sulfinyl group in the stereochemical behavior of **42a** and **42b** as dienophiles was less important than that of configuration at C-5, mainly in reactions conducted in the absence of catalysts [50].

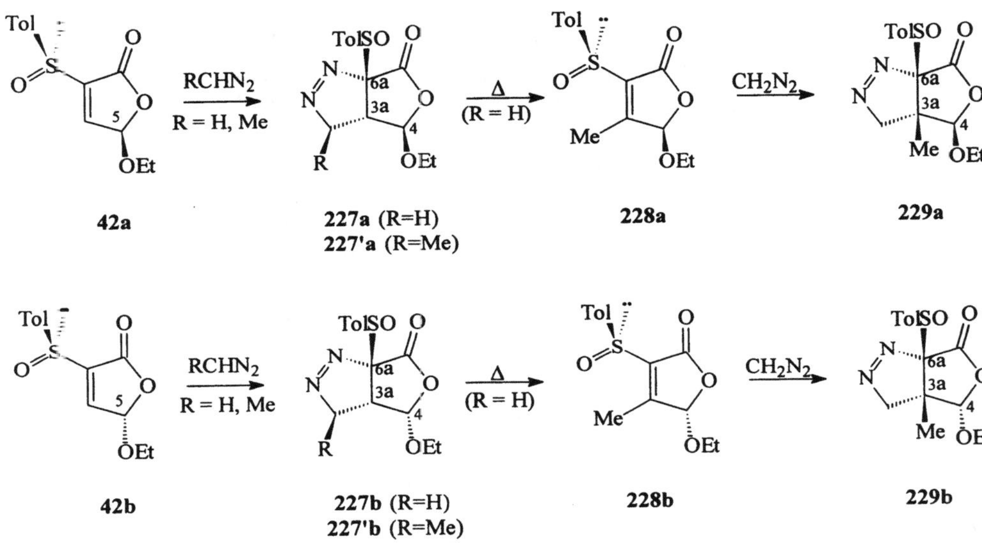

Scheme 104

The extrusion of nitrogen by pyrolysis from **227a** and **227b** yielded 4-methyl sulfinyl butenolides **228a** and **228b**. These were also used as dipolarophiles, and reacted with CH_2N_2 with complete control of regioselectivity and stereoselectivity, affording exclusively pyrazolines **229a** and **229b** (Scheme 104). The stereochemical course of these reactions was explained by assuming steric approach control of diazomethane toward the less hindered face of the dipolarophiles, that bearing the sulfinyl group, in the *s-cis* arrangement.

Results obtained from reactions of (*Z*)-3-*p*-tolylsulfinyl acrylonitriles **58a–c** (Scheme 32) with diazoalkanes [172b] confirmed the conclusions established from the study of **42a** and **42b**.

5
Cycloaditions Mediated by Remote Sulfinyl Groups

In this section are included the results obtained by several authors concerning those asymmetric cycloadditions of substrates different from sulfinyl ethylenes or sulfinyl dienes, but where the stereochemical course is mediated by sulfoxides. This functional group is covalently bonded to one of the reagents, but it is not directly bonded to the double or triple bond acting as a diene, dienophile, or dipolarophile. In order to reach high stereoselectivity, the presence of Lewis acids able to form chelated species involving the sulfinyl oxygen and other heteroatoms present in the sulfinylated reagent are usually necessary.

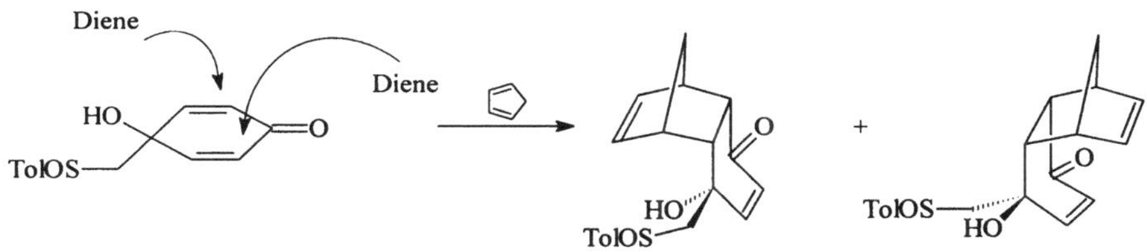

230		231a	231b
	25°C/SiO_2	48	52
	-78°C/BF_3OEt_2	76	24

Scheme 105

The first example illustrating the ability of a sulfinyl group remote to the reactive double bond to control the stereoselectivity of the cycloadditions was the reaction of enantiomerically pure compound **230** (obtained from 4,4-dimethoxy-2,5-cyclohexadienone with (*R*)-methyl-*p*-tolyl sulfoxide) [173]. Mixtures of only two adducts, **231a** and **231b**, were obtained under thermal and catalytic conditions (Scheme 105). Both adducts result by attack of the cyclopentadiene to each of the two endocyclic double bond of **230** from its C-4 hydroxyl face. This high π-facial selectivity was explained on steric grounds, the size of the CH_2SOTol group being larger than that of the OH (stereoelectronic factors were also invoked), and is in agreement with previous results obtained for 4,4-disubstituted cyclohexadienones. The desymmetrization of the prochiral dienone moiety was achieved in the presence of $BF_3.OEt_2$ (2 equiv.), yielding a 3:1 mixture of **231a** and **231b**, easily separated. These open a ready access to diastereoisomerically pure adducts resulting from chiral equivalents of methyl quinols, not easily accessible in other ways. No explanation about the origin of this selectivity is given in this paper. This reaction was further extended to other sulfinylquinones and dienes with very similar stereochemical results. Ab initio calculations provide data on transition states energies for cycloadditions with cyclopentadienes in full agreement with the experimental results [174].

2-*p*-Tolylsulfinyl pyrrole has been recently used as a recoverable chiral auxiliary [175] (Scheme 106). It reacts with unsaturated acyl chlorides yielding **232**

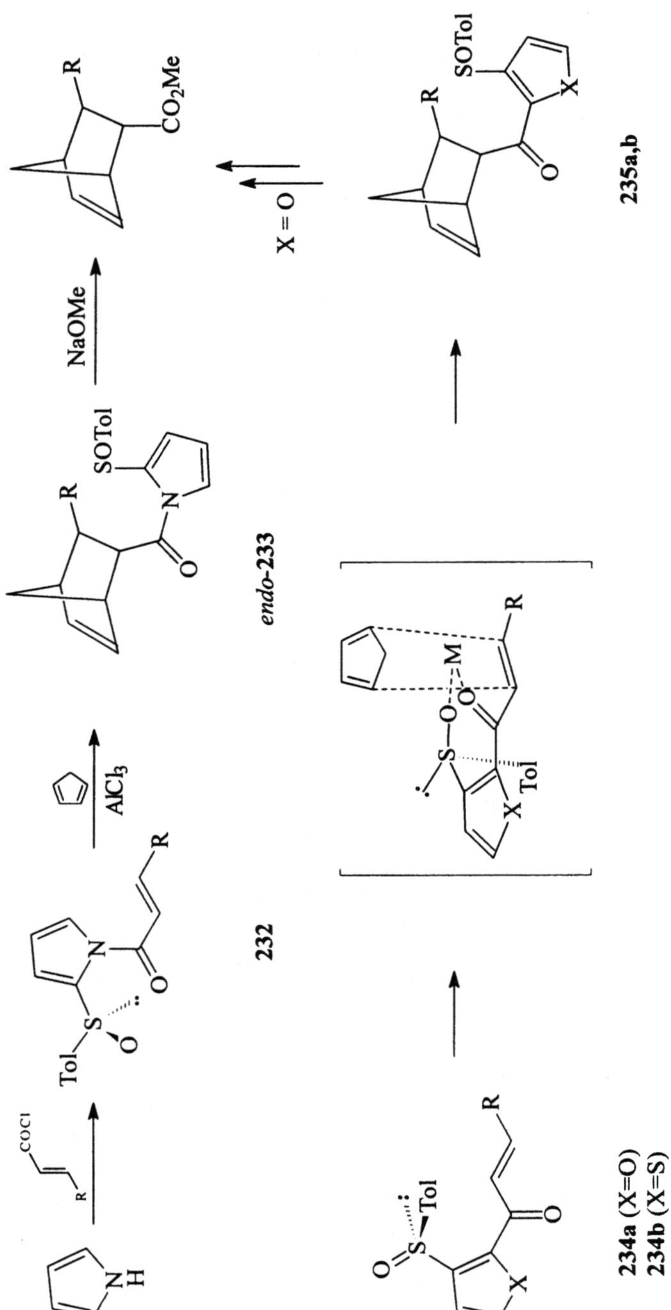
Tol
S
O
N
H
COCl
R
232
AlCl3
endo-233
SOTol
NaOMe
CO2Me
X = O
234a (X=O)
234b (X=S)
M
Tol
X
235a,b

Scheme 106

(R = Ph, Me), which in the presence of cyclopentadiene give a mixture of the four possible diastereoisomers. When reactions were catalyzed by $Yb(OTf)_3$ or $AlCl_3$, the *endo*-**233** adduct was obtained in excellent yield and very high *endo* selectivity (*endo/exo* = 95/5) and π-facial selectivity (*de* ranging from 92% to 99%). Removal of the chiral auxiliary from the adduct could be easily accomplished in quantitative yield with NaOMe, without loss of optical purity (*ee* > 99%). The formation of a chelated species involving the two oxygens of the substrate was invoked to explain the stereochemical results.

Similar results were obtained from reactions of 2-(3-*p*-tolylsulfinyl)-furyl α,β-unsaturated ketones **234a** [175a,b] (R = Me, Ph) and its corresponding 2-thienyl derivative **234b** [175c] (R = Ph) with cyclopentadiene. In the presence of different Lewis acids (with $AlCl_3$ and $Sm(OTf)_3$ being the most efficient), the reactions proceed smoothly to give the *endo*-adducts **235** as the major products in excellent yields and high diastereoselectivity. The *endo/exo* ratio ranges between 90:10 and 96:4 and the π-facial selectivity for the *endo* approach is usually higher than 90%. The stereochemical course of these reactions was rationalized by assuming the formation of different chelated species, one of which (that involving chelation of the sulfinyl and carbonyl oxygens with the metal) is depicted in Scheme 106. As can be seen, the tolyl group seriously hindered the approach of dienes to the lower face of this species. The furan moiety of **235a** was removed by oxidative degradation with RuO_4.

A remote sulfinyl group has also been used to control the stereoselectivity of the hetero-Diels-Alder reaction of the carbonyl group of furfural [176]. Reaction of sulfoxide **236** with Danishefsky's diene in the presence of $Ln(OTf)_3$ (Ln = Yb, Nd, and Sm) yielded cycloadducts **237a** and **237b** with high *de* (93–99%). When reactions were conducted under $Eu(thd)_3$ catalysis, the stereoselectivity of the reaction was dramatically inverted (Scheme 107). The influence of the catalysts in the stereoselectivity is not discussed.

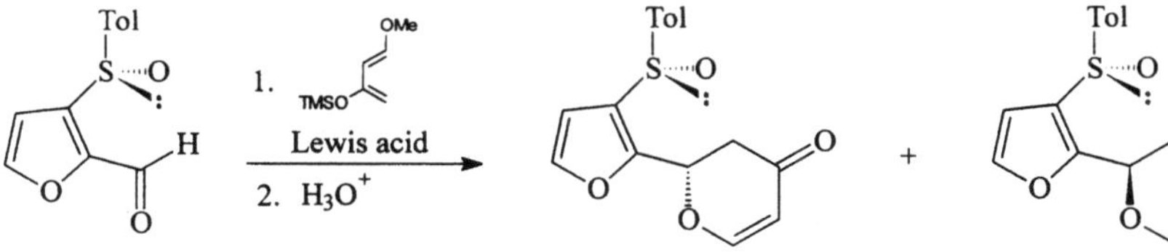

236 **237a** **237b**

catalyst	237a	237b
$Ln(OTf)_3$	>96	<4
$Eu(thd)_3$	12	88

Scheme 107

Asymmetric intramolecular hetero-Diels-Alder reactions mediated by a sulfinyl group not directly bonded to π systems involved in the cycloadditions have also been reported [177]. Thus, Lewis acid catalysed intramolecular cyclization of α'-sulfinyl-α,β-unsaturated ketone, shown in Scheme 108, gave mixtures of

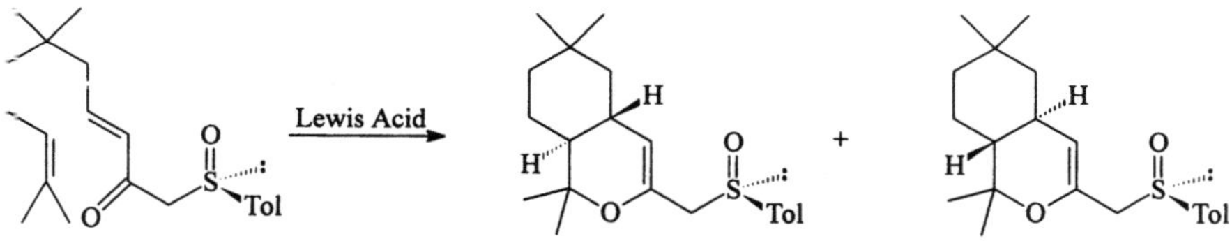

238a 238b

	238a	238b	
$SnCl_4$ (-78°C, 1h)	80	20	(81%)
BF_3OEt_2 (0°C, 2h)	42	58	(74%)
Me_3Al (-78°C, 1h)	40	60	(61%)

Scheme 108

238a and **238b**. The stereochemical course of the reaction depends upon the Lewis acid used. The highest stereoselectivity (*de*=60%) was achieved with $SnCl_4$ and the π-facial diastereoselectivity was moderate or low in all cases.

Some of the problems inherent in 1,3-dipolar cycloadditions of vinylsulfoxides with nitrile oxides, such as their poor regioselectivity and the easy desulfinylation of the adducts, were solved recently by Page et al. They reported cycloaddition of nitrile oxides to α,β-unsaturated acyl dithiane oxides *syn*-**239** and *anti*-**239** [178]. The reactions proved to be remarkably regioselective, only 5-acyl dihydroisoxazoles **240a** and **240b** being isolated (Scheme 109). Compounds *syn*-**239** tended to favor formation of isomer *syn*-**240a** and the induced stereoselectivities were not higher than 5:1 with any of the nitrile oxides used, whereas *anti* substrates tended to favor formation of isomer *anti*-**240b**, the induced stereoselectivities being lower than 3:1. These results suggest that the configuration at C-α at the dithiane unit (but not that at sulfur) is the main controller of the stereochemical course. The dithioacetal moiety may be readily removed by hydrolysis without affecting the dihydroisoxazole rings to provide functionalized dihydroisoxazoles with 1,2-diketone substituents. Additionally, cycloadducts

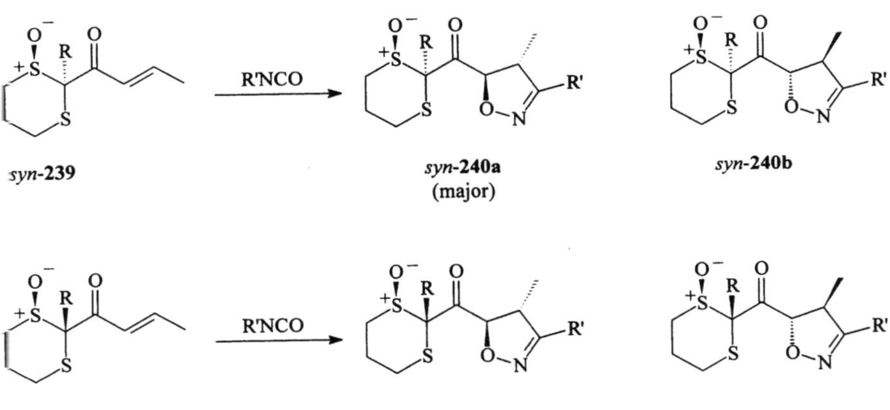

syn-**239** *syn*-**240a** (major) *syn*-**240b**

anti-**239** *anti*-**240a** *anti*-**240b** (major)

Scheme 109

240 could be stereoselectively reduced with l-selectride to provide a third new asymmetric center, as a prior step to the hydrolysis of the dithioacetal unit.

The addition of Lewis acids such as ZnX_2 can alter and even reverse (in reactions starting from *syn*-**239**) the sense of the induced stereoselectivity [179]. The formation of a chelated species in the presence of a Lewis acid with a geometry different to that existing in the absence of catalyst, was invoked to explain the inversion of the stereoselectivity, which remains moderate or low.

Reactions of 3,5-dichloro-2,4,6-trimethyl benzonitrile oxide **241** with fluoromethyl substituted alkenes **242**, bearing a chiral sulfinyl group at β-position of the double bond, afford diastereoisomeric 4,5-dihydroisoxazoles **243** and **244** [180] with a stereoselectivity lower than 2:1 (Scheme 110). The authors conclude that the efficiency of allyl sulfoxides to control diastereoselectivity of 1,3-dipolar cycloadditions with nitrile oxides is lower than that of vinyl sulfoxides.

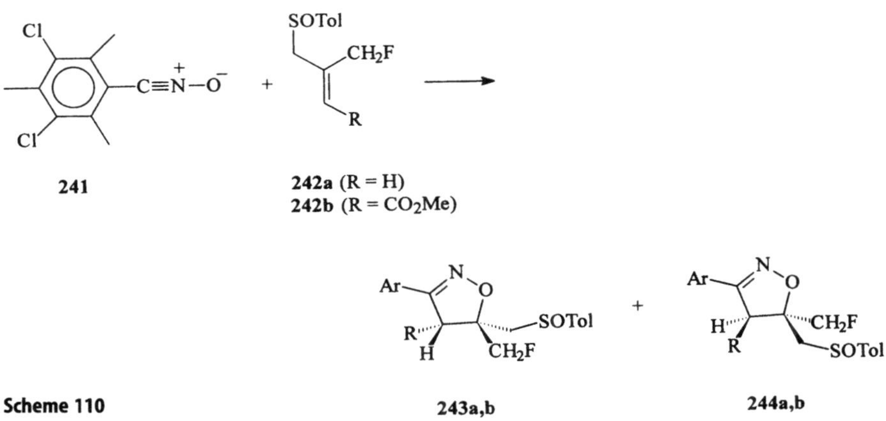

Scheme 110

Finally, enantiomerically pure sulfinimines have also been used as precursors of chiral imidazolidines by 1,3-dipolar cycloaddition with azomethine ylids [181]. Reactions of different arylsulfinimines **245** with dipoles **246** are highly stereoselective, mainly affording diastereoisomer **247** (absolute configuration unequivocally established by X-ray studies), which was readily transformed into vicinal diamine **248** (Scheme 111).

The stereochemical outcome of these reactions may be understood on the basis of the predominant *exo* sulfinyl approach of the ylid (*endo* with respect to the Ar group) toward the less hindered face of the sulfinimine, in s-*cis* conformation. The conformational preferences of the sulfinylimines, which would be responsible for the π-facial selectivity, can be easily rationalized considering the electrostatic repulsion between the lone electron pair at nitrogen and the sulfinyl oxygen, which suggests a great potential interest of these substrates as heterodienophiles and dipolarophiles.

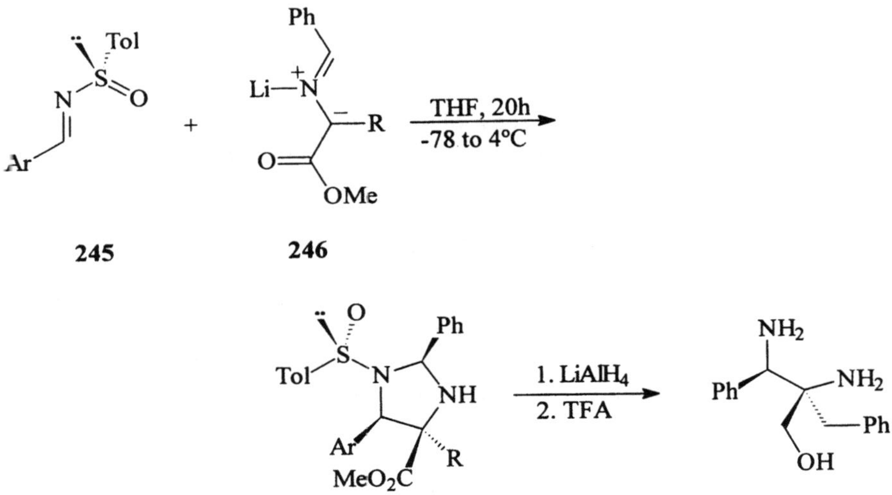

Scheme 111

6 Conclusions

The sulfinyl group has been widely used in asymmetric synthesis to achieve an efficient control of the π-facial selectivity of different types of cycloadditions of vinyl or dienyl sulfoxides. All authors agree that its success is due mainly to the large steric and stereoelectronic differences induced by sulfinyl group on the diastereotopic faces of the neighboring double bonds. It is a consequence of the high *conformational polarizability* of these substrates around the C-S bond, which means that their conformational equilibrium are easily shifted toward some of the possible rotamers.

Discrepancies between different researchers derive from the character inter- or intramolecular of the interactions presumably controlling the reactive conformation. Thus, in most of the cases, the population of the different rotamers in the sulfinylated substrate (only governed by intramolecular interactions) is the only factor considered for explaining the observed π-facial selectivity. This explanation (*static conformational polarization*) was formulated by Koizumi and used by many authors to justify the behavior of vinyl sulfoxides acting as dienophiles and dipolarophiles. A second explanation assumes that the interactions of the two reagents in the transition states determine a different reactivity of the rotamers around the C-S bond. This intermolecular factor can become the most important one in the control of the π-facial selectivity of the cycloadditions, and therefore the tendency expected from conformational stability criteria was not observed in those cases where the most reactive conformation is not the most populated one. This "*dynamic conformational polarization*" has been used just to explain some of the results obtained for sulfinyl quinones and sulfinyl dienes (unexplainable with the above model) but it can be applied to many other cases.

The explanation based on a *"static conformational polarization"* is equivalent to assume that these cycloadditions have *reactant-like TS* (energetic differences in the transition states are similar to those existing between the rotamers of the starting products), whereas the other explanation (*dynamic conformational polarization*) means that they have *product-like TS.* Both explanations could be compatible if we take into account that the position of the transition states on the reaction coordenate must be dependent on the nature of the involved reagents and therefore it has not to be identical in all cases. So far, this question has not been studied.

A second important question relative to the π-facial selectivity of these reactions is the nature of the interactions mainly determining their favored stereochemical course. Although most of the authors assume that the steric effects are principally responsible for the π-facial selectivity of the cycloadditions, some experimental results suggest that electronic factors, acting in the same sense as the steric ones, could also have some influence on the stereoselectivity (see sulfinyl quinones). Nevertheless, additional experimental evidence, not always easy to design, and a theoretical support so far not available, would be necessary to check the last hypothesis.

The use of chelating Lewis acids as catalysts usually improves the stereoselectivity. This influence is mainly important when substrates bear functional groups at a suitable position to become chelated. The metal of the catalyst acts as a template to maintain the complexation of such functional groups and the sulfinyl oxygen. When both are contained in the same reactant, chelation fixes one of the conformations, which in some cases is not the same as that favored in the absence of a catalyst, thus reversing the sense of the π-facial selectivity. This has been the most widely accepted explanation of the influence of these catalysts on the π-facial selectivity of the reactions of substituted vinyl sulfoxides as dienophiles and dipolarophiles. The stereochemical course of some Diels-Alder reactions of sulfinyl dienes with acrylates has been alternatively explained by assuming that the chelating catalyst associates both with the sulfinyl oxygen at the diene and the ester group at the dienophile. The formation of such a species as a prior step to cycloaddition suggests that this must be considered to be an intramolecular process. The steric interactions at the chelated species containing both reagents (not only those existing at the diene) determine the favored stereochemical pathway. These interactions have never been considered to explain the course of the reactions of alkenyl sulfoxides with dienes, but they should be suitable to be taken into account in some cases.

The main problems related to the use of vinyl or dienyl sulfoxides in cycloadditions can be grouped into two different categories, synthetic and mechanistic aspects respectively. Availability of the starting materials, stability of the obtained cycloadducts, and final elimination of the sulfinyl group, are the three main problems to be solved from a synthetic perspective. Reactivity of the substrates and the *endo/exo* selectivity of the cycloadditions will remain as important questions to be answered from a mechanistic viewpoint.

With regard to the preparation of the substrates, which has not been thoroughly discussed here, we will only remark that the efficiency of the synthetic methods so far reported for preparation of some of the most widely used dienophiles, such as sulfinyl acrylates, is not high, low yields often being obtained. Therefore, additional efforts to find a general and successful procedure must be made.

The stability of the adducts is one of the main problems limiting the usefulness of vinyl sulfoxides in asymmetric cycloadditions. Their usually moderate reactivity as dienophiles or dipolarophiles must be compensated by using high temperatures or pressures, or by incorporating other activating groups into the double bond. Both factors favor elimination of the sulfinyl group from the cycloadducts, and this behavior has been utilized in cases where vinyl sulfoxides were used as synthetic equivalents of more unsaturated dienophiles including ketenes, acetylenes, and allenes. Nevertheless, desulfinylation is not desired in asymmetric synthesis because it eliminates two of the four chiral centers created in the cycloaddition. This problem is mainly important in Diels-Alder reactions of vinyl sulfoxides with acyclic dienes (desulfinylation is less favored in bicyclic adducts, resulting from cyclic dienes). In this case, the resulting cyclohexadienes are liable to undergo aromatization, with loss of all of their chiral centers. This is perhaps the main reason that reactions of most sulfinyl dienophiles have also been investigated with cyclopentadiene. Further, cycloadducts resulting from 1,3-dipolar reactions of vinylsulfoxides with dipoles such as nitrile oxides are even less stable, because desulfinylated products are aromatic. Careful design of the dienophiles may be advisable in order to solve these problems. The current use of substrates incorporating the sulfinyl group at remote positions to the reaction centers minimizes or even avoids these problems, but the control of the stereoselectivity is usually less efficient.

Once cycloaddition has been successfully accomplished, elimination of the sulfinyl group must be carried out. Most of the methods used for such a purpose (pyrolytic elimination or hydrogenolysis) do not preserve the configurational integrity of the sulfinylated carbon and thus they give achiral sulfenic acid derivatives. Nevertheless, the main handicap of the sulfinyl group acting as a chiral auxiliary derives from the fact that it does not usually allow the recovering of the original source of chirality, and therefore it does not meet one of the requirements of an ideal chiral inductor. Despite this formal handicap, the high stereoselectivity of the reactions and the moderate costs of the starting materials used as a source of chirality justify continued research in this field. Otherwise, the chemical versatility of the sulfinyl group can be used to transform cycloadducts in other interesting substrates difficult to obtain by other means, in order to compensate for the above handicap. In this sense, we will mention tandem reactions of 1-sulfinyldienes (Diels-Alder/sulfoxide-sulfenate rearrangement), which add interest to these substrates despite their low reactivity. The search for new efficient methods to eliminate the sulfinyl group, mainly those preserving the configurational integrity of the carbon bonded to the sulfur, would widen the synthetic scope of cycloadditions involving vinyl sulfoxides.

From a mechanistic perspective, we think that one of the most striking aspects of these reactions is the influence of the sulfinyl group on the reactivity of the sulfinylated substrates. According to Hammet σ-values, the SOR group must be considered as an efficient electron-withdrawing group. Nevertheless, its influence in dienophilic reactivity is rather low. It can be inferred from the fact that conditions required by substituted sulfinyl ethylenes to react with cyclopentadiene are not much different to those used with the corresponding desulfinylated dienophiles. Moreover, it was clearly deduced from the results obtained in reactions of 2-sulfinyl-*p*-benzoquinone with cyclic and acyclic dienes. The increase in dienophilic reactivity was also evident from the behavior of acyclic dienes, which prefer to react at the sulfinylated C(2)-C(3) double bond, but the rather small extent of this effect in this case was inferred from the results obtained in reactions with cyclic dienes, which preferably react at the non-sulfinylated C(5)-C(6) double bond. These results suggest that steric interactions between the sulfinyl group and the methylenic bridge of the diene are strong enough to overcome the increase in the dienophilicity induced by the sulfinyl group.

In order to explain this small influence of the sulfinyl group increasing the dienophilic reactivity, it must be assumed that the contribution of both of the two different electronic effects of the sulfinyl group (–I and +M) must be significant. Taking into account that the +M effect would mainly increase the energetic contents of the empty LUMO orbital, it would have a negative influence in Diels-Alder reactions (controlled by the interaction $HOMO_{diene}$-$LUMO_{dienophile}$) when the sulfinyl group was bonded to dienophile, but it would be scarcely significant when bonded to diene. In contrast, the –I effect, which mainly affects occupied HOMO orbitals, would explain the low reactivity of sulfinyl dienes.

There are a number of experimental results supporting this proposal. All of them are related to the factors able to modify the extent of the +M effect. Hence, the similar reactivity of sulfinyl acrylates, sulfinyl maleates, and trialkoxycarbonyl sulfinyl ethylenes was explained by assuming that the sulfinyl group could act like a damper for the electronic density, thus enhancing its +M effect as the number of electron-withdrawing substituents increases. Additionally, the reactivity of sulfinylated double bonds is higher when there are structural restrictions preventing the lone electron pair at sulfur from reaching the spatial arrangement required to conjugate with the π-system. This would be the case in alkynyl sulfoxides, which exhibit a clearly higher reactivity than the corresponding alkenyl sulfoxides. In the first substrates, only one of their multiple bonds would be involved in the conjugation which decreases its energy, whereas the second one would be affected only by the –I effect of the sulfinyl group, increasing its reactivity as a dienophile. Nevertheless, the high reactivity of compounds bearing the sulfinyl group contained in a rigid cyclic structure, which restricts the ability to adopt the most suitable spatial arrangement to allow the sulfinyl group to become an electron-donating group, constitutes the main evidence of the important role of the +M effect in the reactivity. In this connection we must point out the different reactivity of the cyclic and acyclic bis-sulfoxides discussed in Sect. 2.1.2.6.

The last problem associated with the use of the sulfinyl group as a chiral inductor in cycloaddition reactions is related to the *endo*/*exo* reactivity. From the available data, the *endo* directing power of the sulfinyl group is not large. The need to incorporate other functional groups into vinyl sulfoxides, thus increasing their dienophilic reactivity, decrees that the formation of mixtures is difficult to avoid in many cases as a consequence of the competition between the groups. Although this factor restricts the usefulness of the vinyl sulfoxides as dienophiles, the usually easy separation of the *endo* and *exo* adducts contributes to minimizing the problem. In the case of the 1,3-dipolar reactions, an *exo* orientating character of the sulfinyl group seemingly may be deduced from the few available results with nitrones. Nevertheless, the confirmation of this assumption requires additional studies.

As a summary, considerable effort is required in order to clarify the role of the sulfinyl group in the reactivity of the substrates. Theoretical calculations would allow determination of the dependence between the orientation of the sulfinyl group and the relative energy of the orbitals of alkenyl and alkynyl sulfoxides and possible influence on the reactivity of Diels-Alder and 1,3-dipolar reactions. These studies should be extended to the transition states involved in cycloadditions in order to clarify the role of the electronic effects on the π-facial selectivity. From a synthetic point of view, the design of sulfoxides minimizing the problems derived from the low dienophilic reactivity and moderate *endo/exo* selectivity of vinyl sulfoxides, and the search of catalysts able to overcome both problems, would increase the value of alkenyl sulfoxides in cycloadditions. Additionally, the use of 1-sulfinyl dienes in hetero-Diels-Alder reactions, and the search for new reactions involving the sulfinyl group at cycloadducts, must form the main targets in future research on sulfinyl dienes. Finally, the least explored field is that of 1,3-dipolar cycloadditions. Here, the design of sulfinyl dipolarophiles able to avoid desulfinylation from their cycloadducts, would allow access to many different types of heterocycles. Regarding mechanistic considerations of these reactions, the role of the sulfinyl group, which seems to be different to that played in Diels-Alder reactions, has not been sufficiently studied, and therefore requires further consideration.

Acknowledgments. We thank Dirección General de Investigación Científica y Técnica (DGICYT), Ministerio de Educación y Cultura (SPAIN) for financial support. We are grateful to Dr. Ana M. Martín Castro and M. Teresa Aranda for their help in preparing the manuscript.

7
References

1. (a) Oppolzer W (1984) Angew Chem Int Ed Engl 23:876; (b)Masamune S, Choy W, Peterson JS, Sita LR (1985) Angew Chem Int Ed Engl 24:1; (c) Helmchen G, Weetman J (1986) In: Sheffold R (ed) Modern synthetic methods, vol 4. Springer, Berlin Heidelberg New York, p 262; (d) Chapuis C, Jurczak (1987) J Hel Chim Acta 70:436; (e) Fringuelli F, Taitchi A, Wenkert E (1990) Org Prep Proc Int 22:131–165; (f) Kagan HB, Riant O (1992) Chem Rev 92:1007; (g) Oh T, Reilly M (1994) Org Prep Proc Int 26:129–158
2. Mislow K, Siegel J (1984) J Am Chem Soc 106:3319 (and references cited therein)
3. (a) Andersen KK (1988) In: Patai S, Rappoport Z, Stirling CJM (eds) The chemistry of sulfones and sulfoxides. Wiley, New York, chap 3, pp 56–94; (b) Walker AJ (1992) Tetrahedron: Asymmetry 3:961–998; (c) Solladié G (1991) In: Trost BM Fleming I (eds) Comprehensive organic synthesis, vol 6. Pergamon, Oxford, chap 3, pp 148–170; (d); Kresze G (1985) In: Klamann D (ed) Methoden der Organischen Chemie (Houben-Weyl). Thieme, Stutgart, pp 669–886

4. (a) Solladié G (1983) In: Morrison JD (ed) Asymmetric synthesis, vol 2. Academic Press, New York, pp 157–199; (b) Posner GH (1983) In: Morrison JD (ed) Asymmetric synthesis, vol 2. Academic Press, New York, pp 225–241; (c) Cinquini M, Cozzi F, Montanari F (1985) In: Bernardi F, Csizmadía G, Magnini A (eds) Organic sulfur chemistry theoretical and experimental advances. Elsevier, Amsterdan, pp 355–408; (d) Posner GH (1987) Acc Chem Res 20:72; (e) Posner GH (1988) In: Patai S Rappoport Z Stirling CJM (eds) The chemistry of sulfones and sulfoxides. Wiley, New York, chap 16, pp 823–849; (f) Hua DH (1992) In: Snieckus V (ed) Advances in the carbanion chemistry, vol 1. JAI Press, London, p 249
5. Maignan C, Raphael RA (1983) Tetrahedron 39:3245
6. Carreño MC (1995) Chem Rev 95:1717–1760
7. Solladie G, Carreño MC (1995) In: Page PCB (ed) Organosulphur chemistry synthetic aspects. Academic Press, New York, chap 1, pp 1–47
8. Jurczak J, Bauer T, Chaouis C (1995) In: Helmchem G, Hoffmann RW, Mulzer J, Schaumann E (eds) Methods of organic chemistry (Houben-Weyl), vol E21b. Georg Thieme, Stuttgart New York, sect 1.6.1, pp 2735–2987
9. De Lucchi O, Pasquato L (1988) Tetrahedron 44:6755–6794
10. (a) Koizumi T (1991) Phosphorus Sulfur and Silicon 58:111–127; (b) Arai Y, Koizumi T (1992) In: Oae S (ed) Reviews on heteroatom chemistry, vol 6. MYU, Tokio, pp 202–217; (c) Arai Y, Koizumi T (1993) Sulfur reports, vol 15. Harwood Academic, pp 41–65
11. Aversa MC, Barattucci A, Bonaccorsi P, Gianneto P (1997) Tetrahedron: Asymmetry 8:1339–1368
12. Paquette LA, Williams RV, Carr RVC, Charumilind P, Blount JF (1982) J Org Chem 47:4566
13. (a) Paquette LA, Moerck RE, Harirchian B, Magnus PD (1978) J Am Chem Soc 100:1597; (b) Miyahara Y, Takahiko I, Yoshino T (1983) Tetrahedron Lett 24:5277; (c) Siegel A (1985) Synthesis 798
14. Williams RV, Chauhan K (1991) J Chem Soc Chem Commun 1672
15. Williams RV, Lin X (1989) J Chem Soc Chem Commun 1872
16. Ji X, van der Helm D, Williams RV, Ebey WJ (1989) Acta Cryst B45:93
17. (a) Kahn SD, Hehre WJ (1986) J Am Chem Soc 108:7399; (b) Kahn SD, Pau CF, Overman LE, Hehre WJ (1986) J Am Chem Soc 108:7381
18. Kahn SD, Hehre WJ (1986) Tetrahedron Lett 27:6041
19. Koizumi T, Arai Y, Takayama H, Kuriyama K, Shiro M (1987) Tetrahedron Lett 28:3689
20. Ronan B, Kagan HB (1991) Tetrahedron: Asymmetry 2:75
21. Ronan B, Kagan HB (1992) Tetrahedron: Asymmetry 3:115
22. Maignan C, Belkasmioui F (1988) Tetrahedron Lett 29:2823
23. Lee AWM, Chan WM, Wong MS (1988) J Chem Soc Chem Commun 1585
24. Maignan C, Belkasmioui F (1989) Bull Soc Chim Fr 695
25. Lee AWM, Chan WM, Ji FY, Poon WH (1995)J Chem Research (S) 368
26. Koizumi T, Hakamada I, Yoshi E (1984) Tetrahedron Lett 25:87
27. Maignan C, Guessous A, Rouessac F (1984) Tetrahedron Lett 25:1727
28. Guessous A, Rouessac F, Maignan C (1986) Bull Soc Chim Fr 837
29. Similar results were obtained in reactions of racemic (*E*)- and (*Z*)-3-phenylsulfinylprop-2-enoic acids and their methyl esters with cyclopentadiene (Proust SM, Ridley DD (1984) Aust J Chem 37:1677) which were investigated to determine the scope of the asymmetric induction of the sulfinyl group It had not been unequivocally ascertained by other research groups working in this area, mainly Montanary's group (the best revision of the pioneer and important contribution of Montanary's group to the study of Diels-Alder reaction with racemic sulfinyl acrylates was made by De Lucchi and Pasquato [9]. The most significant contribution of the paper by Ridley was the negative role of the Lewis acid on the stereoselectivity
30. Brimble MA, Davis BR (1985) Tetrahedron 41:4965
31. (a) Arai Y, Yamamoto M, Koizumi T (1986) Chem Lett 1225; (b) Arai Y, Yamamoto M, Koizumi T (1988) Bull Chem Soc Jpn 61:467
32. Baumann M, Hoffmann W (1979) Liebigs Ann Chem 743

33. (a) De Lucchi O, Lucchini V, Marchioro C, Valle J, Modena G (1986) J Org Chem 51:1457; (b) De Lucchi O, Lucchini V, Marchioro C, Valle J, Modena G (1989) (Corrigenda) J Org Chem 54:3245)
34. Takayama H, Iyobe A, Koizumi T (1986) J Chem Soc Chem Commun 771
35. Arai Y, Hayashi Y, Matatoshi Y, Takayama H, Koizumi T (1987) Chem Lett 185
36. Arai Y, Hayashi Y, Matatoshi Y, Takayama H, Koizumi T (1988) J Chem Soc Perkin Trans I 3133
37. Arita M, Adachi K, Ito Y, Saway H, Ohno M (1983) J Am Chem Soc 105:4049
38. Takayama H, Iyobe A, Koizumi T (1987) Chem Pharm Bull 35:433
39. Takahashi T, Iyobe A, Arai Y, Koizumi T (1989) Synthesis 189
40. Takahashi T, Katsubo H, Iyobe A, Namiki T, Koizumi T (1990) J Chem Soc Perkin Trans I 3065
41. Takayama H, Hayashi K, Takeuchi Y, Koizumi T (1986) Heterocycles 24:2137
42. Takayama H, Hayashi K, Koizumi T (1986) Tetrahedron Lett 27:5509
43. Yamakoshi YN, Ge W-Y, Sugita J, Okayama K, Takahashi T Koizumi T, (1996) Heterocycles 42:129
44. Martinow J, Dimitroff H, Fallis AG (1993) Tetrahedron Lett 34:8201
45. (a) Arai Y, Kuwayama S, Takeuchi Y, Koizumi T (1985) Tetrahedron Lett 26:6205; (b) Chau T-Y, Liu H-C, Hu W (1996) Synthetic Commun 1867
46. Carretero JC, García Ruano JL, Martín Cabrejas LM (1997) Tetrahedron: Asymmetry 8:2215
47. Chau T-Y, Lin H-C, Hu W (1985) Tetrahedron Lett 26:5453
48. Alexandre C, Rouessac F, Tabti B (1996) Synt Commun 26:1867
49. Carretero JC García Ruano JL, Martín Cabrejas LM (1997) Tetrahedron: Asymmetry 8:1623
50. Carretero JC, García Ruano JL, Lorente A, Yuste F (1993) Tetrahedron: Asymmetry 4:177
51. (a) de Jonh JC, Bolhuis F, Feringa BL (1991) Tetrahedron: Asymmetry 2:1247 (and references cited therein); (b) Casas R, Parella T, Brancharell V, Ortuño R (1992) Tetrahedron 48:2659 (and references cited therein)
52. Carretero JC, García Ruano JL, Martín Cabrejas L (1999) Tetrahedron: Asymmetry (in press)
53. Maignan C, Guessous A, Rouessac F (1986) Tetrahedron Lett 27:2603
54. Alcudia F, Llera JM, Ordoñez M (1995) Tetrahedron Lett 36:4889
55. Ordoñez M (1996) PhD Dissertation, Universidad de Sevilla
56. (a) Alexandre C, Belkadi O, Maignan C (1992) J Chem Research 48; (b) Alexandre C, Goesdon A, Rouessac F (1997) Bull Soc Chim France 719
57. Nishio T, Tokunaga T, Omote Y (1988) Synth Commun 18:2083
58. Alonso I, Carretero JC, García Ruano JL (1989) Tetrahedron Lett 30:3853
59. Alonso I, Carretero JC, García Ruano JL, Martín LM, Lopez-Solera I, Raithby PR (1994) Tetrahedron Lett 35:9461
60. Danishefsky S, Morris J, Clizbe LA (1981) J Am Chem Soc 103:1602
61. Danishefsky S (1981) Acc Chem Res 14:400 (and references cited therein)
62. Alonso I, Carreño MC, Carretero JC, García Ruano JL, Urbano AR (1990) Acad Ciencias Exactas Fis y Natur 84:167
63. Weiler ED, Brennan JJ (1978) J Heterocyclic Chem 15:1299
64. Waldner A (1989) Tetrahedron Lett 30:3061
65. García Ruano JL, Esteban A, Martín Castro AM, Rodriguez Ramos JH, Solera MI (1998) J Org Chem 63:3324
66. Mikolajczyk M, Midura WH (1992) Tetrahedron: Asymmetry 3:1515
67. Lopez R, Carretero JC (1991) Tetrahedron: Asymmetry 2:93
68. Redon M, Janousek Z, Viehe HG (1997) Tetrahedron 53:6861
69. (a) Ono N, Kamimura K, Kaji A (1986) J Org Chem 51:2139; (b) Ono N, Kamimura K, Kaji A (1988) J Org Chem 53:251
70. Yung M, Grove DD (1987) J Chem Soc Chem Commun 753

71. Fuji K, Tonaka K, Abe H, Itoh A, Node M, Taga T, Miwa Y, Shiro M (1991) Tetrahedron: Asymmetry 2:179
72. Fuji K, Tonaka K, Abe H, Itoh A, Node M, Taga T, Miwa Y, Shiro M (1991) Tetrahedron: Asymmetry 2:1319
73. Fuji K, Tonaka K, Abe H, Matsumoto K, Taga T, Miwa Y (1992) Tetrahedron: Asymmetry 3:609
74. Fuji K, Tonaka K, Abe H, Matsumoto K, Harayama T, Ikeda A, Taga T, Miwa Y, Node M (1994) J Org Chem 59:2211. The use of $ZnCl_2$ to catalyze the reaction of **70** with cyclopentadiene has been recently reported (see Abe H, Fujii H, Masunari C, Itani J, Kashino S, Shibaike K, Harayama T (1997) Chem Pharm Bull 45:778)
75. (a) Schaumann E (1985) In: Klamann D (ed) Methoden der Organischen Chemie (Houben-Weyl) Georg Thieme Verlag Stutgart Band E-11, pp 260–324; (b) Kolb M (1988) In: Patai S (ed) The chemistry of functional groups: the chemistry of ketenes, allenes and related compounds. Wiley Interscience, chap 16, pp 669–699; (c) Harirchian B, Magnus P (1977) J Chem Soc Chem Commun 522; (d) Moss WO, Bradbury RH, Hales NJ, Gallagher T (1988) Tetrahedron Lett 29:6475; (e) Hayashi Y, Narasaka K (1989) Chem Lett 793
76. (a) Herrmann JL, Kieczykowski GR, Pomanet RF, Wepplo PJ, Schlessinger RH (1973) Tetrahedron Lett 4711; (b) Herrmann JL, Kieczykowski GR, Pomanet RF, Schlessinger RH (1973) Tetrahedron Lett 4715; (c) Ban Y, Ohnuma T, Seki K, Oishi T (1975) Tetrahedron Lett 727; (d) Kieczykowski GR, Pogonowski CS, Richman JE, Schlessinger RH (1977) J Org Chem 42 175; (e) Davis R, Untch KG (1979) J Org Chem 44:3755
77. (a) De Lucchi O, Lucchini V, Pasquato L, Zamai M, Modena G (1984) Gazz Chim Ital 114:293; (b) De Lucchi O, Pasquato L, Modena G (1984) Tetrahedron Lett 25:3643; (c) Koteswar Rao Y, Nagarajan M (1984) Synthesis 757; (d) Neplyuev VM, Bazarova IM, Lozinskii MO (1986) Russ Chem Rev 55:1550; (e) Himbert G, Kosack S (1988) Chem Ber 121:2053
78. Arai Y, Kuwayama S, Takeuchi Y, Koizumi T (1986) Synth Commun 16:233
79. De Lucci O, Fabbri D, Lucchini V (1991) Synlett 565
80. Aggarwal VN, Lightowler M, Lindell SD (1992) Synlett 730
81. Aggarwal VN, Dravowicz J, Grainger RS, Gültekin Z, Lightowler M, Spargo PL (1995) J Org Chem 60:4962
82. Katagiri N, Ise S, Watanabe N, Kaneko C (1990) Chem Pharm Bull 38:3242
83. Carretero JC, García Ruano JL, Martín LM (1997) Tetrahedron: Asymmetry 8:409
84. (a) Bertotti E, Luciani G, Montanari F (1959) Gazzetta Chim It 89:1564; (b) Cinquini M, Colonna S, Montanari F (1970) J Chem Soc 572; (c) Cecchet E, Di Furia F, Licini G, Modena G (1996) Tetrahedron: Asymmetry 7:369
85. Kaydos JA, Smith DL (1983) J Org Chem 48:1096
86. (a) Arai Y, Hayashi K, Koizumi T, Shiro M, Kuriyama K (1988) Tetrahedron Lett 29:6143; (b) Arai Y, Matsui M, Koizumi T (1990) J Chem Soc Perkin Trans I 1233
87. Arai Y, Matsui M, Koizumi T (1990) Chem Lett 1585
88. Arai Y, Hayashi K, Matsui M, Koizumi T, Shiro M, Kuriyama KJ (1991) Chem Soc Perkin Trans I 1709
89. Alonso I, Carretero JC, García Ruano JL (1991) Tetrahedron Lett 32 947
90. Alonso I, Cid B, Carretero JC, García Ruano JL, Hoyos MA (1991) Tetrahedron: Asymmetry 2:1193
91. Alonso I, Carretero JC, García Ruano JL (1993) J Org Chem 58:3231
92. Alonso I, Carretero JC, García Ruano JL (1994) J Org Chem 59:1499
93. Carretero JC, García Ruano JL, Martín LM (1994) Tetrahedron Lett 35:5895
94. Carretero JC, García Ruano JL, Martín LM (1995) Tetrahedron 51:8323
95. Arai Y, Matsui M, Koizumi T, Shiro M (1991) J Org Chem 56:1983
96. Arai Y, Matsui M, Fujii A, Kontani T, Ohno T, Koizumi T, Shiro M (1994) J Chem Soc Perkin Trans I 25
97. Arai Y, Fujii A, Ohno T, Koizumi T (1992) Chem Pharm Bull 40:1670
98. Arai Y, Kontani T, Koizumi T (1994) J Chem Soc Perkin Trans I 15

99. Arai Y, Kontani T, Koizumi T (1991) Chem Lett 2135
100. Arai Y, Kontani T, Koizumi T (1992) Tetrahedron: Asymmetry 3:535
101. Boeckman RK Jr, Dolak TM, Culos KO (1978) J Am Chem Soc 100:7098
102. Boeckman RK Jr, Delton MH, Dolak TM, Watanabe T, Glick MD (1979) J Org Chem 44:4396
103. Kraus GA, Woo SH (1986) J Org Chem 51:7098
104. Iwao M, Kuraishi T (1987) Bull Chem Soc Jpn 60:4051
105. Carreño MC, García Ruano JL, Urbano A (1989) Tetrahedron Lett 30:4003
106. Carreño MC, Mahugo J, Urbano A (1997) Tetrahedron Lett 38:3047
107. (a) Carreño MC, García Ruano JL, Urbano A (1992) Synthesis 651; (b) Carreño MC, García Ruano JL, Mata JM, Urbano A (1991) Tetrahedron 47:605
108. Carreño MC, García Ruano JL, Urbano A (1994) Tetrahedron 50:5013
109. Carreño MC, García Ruano JL, Toledo MA, Urbano A (1994) Tetrahedron Lett 35:9759
110. Carreño MC, García Ruano JL, Urbano A (1992) J Org Chem 57:6870
111. Carreño MC, García Ruano JL, Urbano A (1994) Tetrahedron Lett 35:3789
112. García Ruano JL, Carreño MC, Carretero JC, Martín LM, Urbano A (1996) Pure & Appl Chem 68:925
113. Carreño MC, García Ruano JL, Toledo MA, Urbano A, Remor C, Stefani W, Fisher J (1996) J Org Chem 61:503
114. Bartolomé JM, Carreño MC, Urbano A (1996) Tetrahedron Lett 37:3187
115. Carreño MC, García Ruano JL, Urbano A, Hoyos MA (1996) J Org Chem 61:2980
116. Carreño MC, García Ruano JL, Urbano A, Lopez-Solera MI (1997) J Org Chem 62:976
117. (a) Carreño MC, Urbano A, Fisher J (1997) Angew Chem Int Ed Eng 36:1621; (b) Carreño MC, Urbano A, Di Vitta C (1998) J Org Chem 63:8320; (c) Carreño MC, García Cerrada S, Urbano A, Di Vitta (1998) Tetrahedron: Asymmetry 9:2965
118. (a) Carreño MC, García Ruano JL, Remor CZ, Urbano A, Fisher J (1997) Tetrahedron Lett 38:9077; (b) Carreño MC, Hernandez Sanchez R, Mahugo J, Urbano A (1999) J Org Chem (in press); (c) Carreño MC, García Ruano JL, Toledo MA (unpublished results). Preliminary communication was presented in 11th International Conference on Organic Synthesis (ICOS-11) Amsterdam July (1996), Book of Abstracts, p 351
119. (a) Evans DA, Bryan CA, Sims CL (1972) J Am Chem Soc 94:2891; (b) Overman LE, Petty CB, Takashi B, Huang GT (1983) J Am Chem Soc 105:6335; (c) Posner GH, Harrison W, Wettlaufer DG (1985) J Org Chem 50:5041; (d) Fisher MJ, Hehre WJ, Kahn SD, Overman LE (1988) J Am Chem Soc 110:4625; (e) Naperstkow AM, Macaulay JB, Newlands MJ, Fallis AG (1989) Tetrahedron Lett 30:5077; (f) Makaulay JB, Fallis AG (1988) J Am Chem Soc 110:4074; (g) Backval J-E, Ericsson A (1994) J Org Chem 59:5850
120. (a) Fischer MJ, Overman LE (1988) J Org Chem 53:2630; (b) Carey FA, Hernandez O (1973) J Org Chem 38:2670; (c) Trost MB, Bridges AJ (1975) J Org Chem 40:2014
121. (a) Smith G, Stirling CJM (1971) J Chem Soc 1530; (b) Jeganathan S, Okamura WH (1982) Tetrahedron Lett 23:4763; (c) Bridges AJ, Fisher JW (1982) J Chem Soc Chem Commun 665; (d) Ni Z, Wang X, Rodriguez A, Padwa A (1992) Tetrahedron Lett 33:7303
122. Gaoni Y (1977) Tetrahedron Lett 18:4524
123. (a) Gaonac'h O, Maddaluno J, Plè G, Duhamel L (1992) Tetrahedron Lett 33:2473; (b) Yang T-K, Chu H-Y, Lee D-S, Jiang Y-Z, Chou T-S (1996) Tetrahedron Lett 37:4537
124. (a) Hoffmann RW, Goldmann S, Maak N, Geueke K (1978) J Chem Ber 113:831; (b) Solladié G, Ruiz P, Colobert F, Carreño MC, García Ruano JL (1991) Synthesis 1011; (c) Solladié G, Colobert F, Maugein N, Moreno I, Almario A, Carreño MC, García Ruano JL (1992) Tetrahedron Lett 33:4561; (d) Solladié G, Colobert F, Ruiz P, Hamdouchi C, Carreño MC, García Ruano JL (1991) Tetrahedron Lett 32:3695; (e) Carreño MC, Cid MB, Colobert F, García Ruano JL, Solladié G (1995) Tetrahedron: Asymmetry 5:1439; (f) Pindur U, Lutz G, Rogge M (1995) J Heterocyclic Chem 32:201
125. (a) Paley RS, de Dios A, Fernandez de la Pradilla R (1993) Tetrahedron Lett 34:2429; (b) Paley RS, Lafontaine JA, Ventura MP (1993) Tetrahedron Lett 34:3663
126. Arce E, Carreño MC, Cid MB, García Ruano JL (1995) Tetrahedron: Asymmetry 6:1757

127. (a) Aversa MC, Bonoccorsi P, Giannetto P, Jafari SMA, Jones DN (1992) Tetrahedron: Asymmetry 3:701; (b) Aversa MC, Bonoccorsi P, Giannetto P, Jones DN (1994) Tetrahedron: Asymmetry 5:805
128. Gibbs RA, Okamura WH (1988) J Am Chem Soc 110:4062
129. Paley RS, Weers HL, Fernadez P (1995) Tetrahedron Lett 36:3605
130. (a) Bonfand E, Gosselin P, Maignan C (1992) Tetrahedron Lett 33:2347; (b) Bonfand E, Gosselin P, Maignan C (1993) Tetrahedron: Asymmetry 4:1667; (c) Gosselin P, Bonfand E, Maignan C (1996) Synthesis 1079
131. Aranda MT, Aversa MC, Barattucci A, Cid MB, Carreño MC, García Ruano JL (unpublished results). A preliminary communication was presented in Convegno Nazionale Divisione di Chimica Societé Chimica Italiana Monopoli (1996), Book of Abstracts, p 98
132. Ando W, Hanyu Y, Takata T (1982) J Am Chem Soc 104:4981
133. (a) Posner GH, Harrison W (1985) J Chem Soc Chem Commun 1786; (b) Posner GH (1987) Acc Chem Res 20:72
134. Posner GH, Haces A, Harrison W, Kinter CM (1987) J Org Chem 52:4836
135. Cieplak AS (1981) J Am Chem Soc 103:4540
136. Treiber A, Dansette PM, Amri HE, Girault J-P, Ginderow D, Mornon J-P, Mansuy D (1997) J Am Chem Soc 119:1565
137. (a) Adams H, Jones DN, Aversa MC, Bonaccorsi P, Giannetto P (1993) Tetrahedron Lett 34:6481; (b) Aversa MC, Barattucci A, Bonaccorsi P, Giannetto P, Jones DN (1997) J Org Chem 62:4376; (c) Aversa MC, Barattucci A, Bonaccorsi P, Giannetto P, Panzolorto M, Rizzo S (1998) Tetrahedron: Asymmetry 9:1577
138. In the original paper [137a] there is a mistake in Fig. 2 showing the transition state yielding endo A-153a. It has been corrected in Fig. 2 of [137b]
139. Gosselin P, Bonfand E, Hayes P, Retoux R, Maignan C (1994) Tetrahedron: Asymmetry 5:781
140. Pradilla RF, Montero C, Viso A (1998) J Chem Soc Chem Commun 409
141. Gosselin P, Bonfand E, Maignan C (1996) J Org Chem 61:9049
142. Arce E, Carreño MC, Cid MB, García Ruano JL (1994) J Org Chem 59:3421
143. (a) Carreño MC, Cid MB, García Ruano JL (1996) Tetrahedron: Asymmetry 7:2151; (b) Carreño MC, Cid MB, García Ruano JL, Santos M (1997) Tetrahedron: Asymmetry 8:2093
144. Bell AS, Fishwick CWG, Reed JE (1995) Tetrahedron Lett 36:7713
145. For a recent review see Tietze LF, Kettschau J (1997) In: Metz P (ed) Topics Curr Chem 189:4
146. (a) Hiroi K, Masayuki U, Fujisawa A (1992) Tetrahedron Lett 33:7161; (b) Hiroi K, Masayuki U, Fujisawa A (1993) Chem Pharm Bull 41:666
147. Gosselin P, Bonfand E, Hayes P, Retoux R, Maignan C (1994) Tetrahedron: Asymmetry 5:781
148. Hayes P, Dujardin G, Maignan C (1996) Tetrahedron Lett 37:3687
149. Aversa MC, Barattucci A, Bonaccorsi P, Bruno G, Giannetto P, Panzalorto M (1997) Tetrahedron: Asymmetry 8:2989
150. Carreño MC, Cid MB, García Ruano JL, Santos M (1998) Tetrahedron Lett 39:1405
151. (a) Padwa A (ed) 13-Dipolar cycloaddition chemistry (1984) Wiley, New York; (b) Torsell KBG (1988) In: Nitrile oxides nitrones and nitronates in organic synthesis. VCH Heinheim; (c) Kozikowski AP (1984) Acc Chem Res 17:410; (d) Kanemasa S, Tesuge O (1990) Heterocycles 30:71; (e) Gothelf KV, Jorgensen KA (1998) Chem Rev 98:863
152. Houk KN, Firestone RA, Munchausen LL, Mueller PH, Arison BH, García LA (1985) J Am Chem Soc 107:7227
153. Koizumi T, Hirai H, Yoshii E (1982) J Org Chem 47:4004
154. Takahashi T, Fujii A, Sugita J, Hagi T, Kitano K, Arai Y, Koizumi T (1991) Tetrahedron: Asymmetry 2:1379
155. Mullen GB, Georgiev VS (1989) J Org Chem 54:2476
156. Louis C, Mill S, Mancuso V, Hootelé C (1994) Can J Chem 72:1347

157. (a) Takahashi T, Kitano K, Hagi T, Nikonmatsu H, Koizumi T (1989) Chemistry Lett 597; (b) Araldi GL, Prakash KRC, George C, Kozikowski AP (1997) J Chem Soc Chem Commun 1875
158. Bravo P, Bruché L, Farina A, Fronza G, Meille SV, Merli A (1993) Tetrahedron: Asymmetry 4:2131
159. (a) Louis C, Hootelé C (1995) Tetrahedron: Asymmetry 6:2149; (b) Louis C, Hootelé C (1997) Tetrahedron: Asymmetry 8:109
150. Hotokka M, Kimmelma R (1994) J Mol Struct (Theochem) 313
151. Aggarwal VK, Grainger RS, Adams H, Spargo PL (1998) J Org Chem 63:3481
152. Bravo P, Ponti PP (1973) J Heterocyclic Chem 10:669
153. Caramella P, Albini E, Bandiera T, Coda AC, Grunanger P, Albini FM (1983) Tetrahedron 39:689
154. Geneste P, Durand R, Pioch D (1979) Tetrahedron Lett 4845
155. Bened A, Durand R, Pioch D, Geneste P, Declercq JP, Germain G, Rambaud J, Roques R, Guimon C, Guillouzo GP (1982) J Org Chem 47:2461
156. (a) Bravo P, Bruché L, Diliddo D, Resnati G (1993) J Chem Research (S) 346; (b) Bravo P, Bruché L, Merli A, Fronza G (1994) Gazz Chim Ital 124:275
157. García Ruano JL, Fraile A, Martín MR (1999) Tetrahedron (submitted)
158. Veniard L, Pourcelot G (1973) Bull Soc Chim France 2746
159. Dean FM, Park BK (1974) Tetrahedron Lett 4275
170. Padwa A, Meske M, Rodriguez A (1995) Heterocycles 40:191
171. Planquaert MA, Redon M, Janousek Z, Viehe HG (1996) Tetrahedron 52:4383
172. (a) García Ruano JL, Fraile A, Martín MR (1996) Tetrahedron: Asymmetry 7:1943; (b) García Ruano JL, Fraile A, Martín MR (unpublished results); (c) Blanco D, García Ruano JL, Martín Castro AM, Martín MR, Rodriguez Ramos JH (1998) 18th International Symposium on the Organic Chemistry of Sulfur Firenze, July 1998
173. Carreño MC, Pérez González M, Fischer J (1995) Tetrahedron Lett 36:4893
174. Carreño MC, Pérez González M, Houk KN (1997) J Org Chem 62:9128
175. (a) Arai Y, Masuda T, Masaki Y (1997) Chem Lett 145; (b) Arai Y, Masuda T, Masaki Y, Shiro M (1996) J Chem Soc Perkin I 759; (c) Arai Y, Masuda T, Masaki Y (1998) Chem Pharm Bull 46:1078
176. Arai Y, Masuda T, Masaki Y, Shiro M (1997) Tetrahedron: Asymmetry 8:1199
177. Hiroi K, Umemura M, Tomikawa Y (1993) Heterocycles 35:73
178. Page PCB, Purdie M, Lathbury D (1997) Tetrahedron 53:1061
179. Page PCB, Purdie M, Lathbury D (1997) Tetrahedron 53:7365
180. Bravo P, Bruché L, Seresini P, Zanda M, Arnone A (1997) J Heterocyclic Chem 34:489
181. Viso A, Fernandez de la Pradilla R, Guerrero-Strachan C, Alonso M, Martinez-Ripoll M, André I (1997) J Org Chem 62:2316

Thiocarbonyl Compounds as Specific Tools for Organic Synthesis

Patrick Metzner

Laboratoire de Chimie Moléculaire et Thio-organique (UMR CNRS # 6507), ISMRA – Université de Caen, 6 Boulevard du Maréchal Juin, F-14050 Caen, France
E-mail: metzner@ismra.fr

Thiocarbonyl compounds (thioamides, thioesters, thioketones...) react readily with a large variety of reagents (nucleophiles, electrophiles and radicals) due to their weak C=S bond and the aptitude of sulfur to stabilise an adjacent charge or radical centre. Thus, nucleophilic additions, deprotonation, and sigmatropic rearrangements are often more facile than in the oxygen series. Moreover, a number of specific reactions have been uncovered: thiophilic addition of nucleophiles, Michael addition of enethiolates, the Eschenmoser reaction, oxidation to sulfines, a large variety of [4+2] and [3+2] cycloaddition reactions with 1,3-dienes and 1,3-dipoles. Far from being purely exotic species, thiocarbonyl compounds are now efficient and specific tools and have indeed been used in multi-step synthetic schemes leading to various products of biological interest.

Keywords: Thiocarbonyl compounds, Thiophilic addition, Cycloaddition reactions, Enethiolisation, Eschenmoser reaction, Sigmatropic rearrangement.

Topics in Current Chemistry, Vol. 204

1 Introduction

Thiocarbonyl compounds have recently emerged as synthetic tools with specific properties. Although some thioamides and thioketones were prepared as early as the 19th century, general methods are rather recent. Nowadays most, if not all, thiocarbonyl compounds that one can imagine can be prepared, with techniques adapted to the stabilities of the target molecules.

A diverse situation is encountered, from *thioamides* which are easy to synthesise in large amounts to highly reactive *thioaldehydes* which are usually prepared in situ in the presence of a reagent. In between, *dithioesters* and *thionoesters* are generally simple to prepare and stable molecules. Thioketones are somewhat tricky molecules with some reasonably stable model molecules to work with. These synthetic efforts, mostly completed in the 1970s and 1980s, have recently allowed more widespread use of thiocarbonyl compounds in terms of the present challenges of organic synthesis: reversal of reactivity (e.g. thiophilic addition, *S*-alkylation of enethiolates), enhanced reactivity (thioacylation, sigmatropic shifts), specific behaviour (Eschenmoser reaction, [4+2] and dipolar cycloadditions, formation of sulfines, desulfurisation of thioamides, fluorination).

The present review concentrates on the developments reported in the last few years (1992 to May 1997) and mostly on carbon-carbon bond forming reactions. Other classes of compounds are generally not discussed, including: thiocarbonates, dithiocarbonates and thiocarbamates. The chemistry of thiocarbonyl *S*-oxides is presented and compared.

In 1992 a related review [1] appeared, which can be consulted for earlier chemistry and other review references. Since then, the following reports have appeared:

- A general and concise monograph by Whitham [2]
- In the "*Best Synthetic Methods*" series a presentation by Thuillier and the present author of a selection of experimental procedures of "*Sulfur Reagents in Organic Synthesis*" [3]

- Thioaldehydes have been reviewed by Okazaki [4], and in a report by Kirby [5]
- The chemistry of thio- and dithioesters has been compiled by Kato and Murai [6]
- The *Comprehensive Organic Functional Group Transformations* series extensively presents the synthesis of various thiocarbonyl compounds [7]

2 Recent Syntheses of Thiocarbonyl Compounds

The efforts of specialists in this field and also those of organic chemists in search of specific transformations have now demonstrated that many thiocarbonyl compounds are not highly unstable molecules and can be prepared without difficulty. The general series which have appeared very recently, such as *Comprehensive Organic Synthesis* [8, 9] or *Encyclopedia of Reagents for Organic Synthesis* (for example [10, 11]) present a good deal of information on their preparation.

The *Comprehensive Organic Functional Group Transformations* series includes up-to-date overviews of the synthesis of thioaldehydes and thioketones [12], thionoesters [13], dithioesters [14], and thioamides [15, 16].

A monograph on the use of flash vacuum thermolysis techniques (FVT) has been edited by Vallée [17] and a specific review on thioketones and thioaldehydes has appeared [18]. Overviews have focused on carbohydrates bearing the thiocarbonyl group [19] and on thiopeptides [20].

We present here the most recent reports, arranged by chemical group: thioamides, dithioesters, thionoesters, thioketones, thioaldehydes and sulfines.

2.1 Thioamides

This class of compounds has attracted a great deal of attention related to their ease of synthesis, excellent stabilities, crystalline form and absence of obnoxious smell. Over the last 4 years, tens of papers have introduced new methods or applications.

One of the oldest methods is still being used: sulfurisation of amides with tetraphosphorus decasulfide (P_4S_{10}). Although it involves heterogeneous reactions, it is often an efficient reagent, as in the synthesis of an intermediate towards *Vinca* alkaloid related compounds [21]. Ultrasonic activation was successful for the preparation of sugar derivatives [22] and thiolactams [23].

A revival of this old thionating agent was proposed by Hartke and Gerber [24]. The combination of P_4S_{10} with sodium fluoride (molar ratio 1:2) in DME gave excellent conversions of amides into thioamides: mild conditions, high yields and easy work-up. The nature of the reactive species is not known.

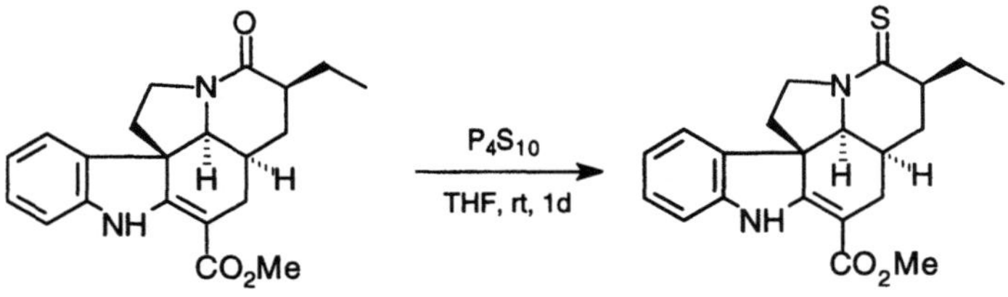

Scheme 1

Since the pioneering work of Lawesson, who developed a phosphorosulfur compound which is reasonably soluble in organic solvents to allow homogeneous and efficient carbonyl thionation, a number of applications have been reported, especially in the context of sulfur analogues of biologically active molecules bearing an amide group [25]. Schaumann and Nieschalk [26] have prepared sulfur analogues of monolactams in moderate yields: the Davy reagent was found preferable to the Lawesson reagent. Using solid-state photochemistry, Sakamato and co-workers [27] have achieved an impressive absolute asymmetric synthesis of a β-thiolactam from an achiral unsaturated thioamide with 81–97% e.e. A thioamide bearing an α-asymmetric centre with an alkoxy group was obtained with a 94% e.e. by using the highly soluble Belleau reagent at 0 °C [28]. In another example, Heathcock and Sharp report an unusual epimerisation [29].

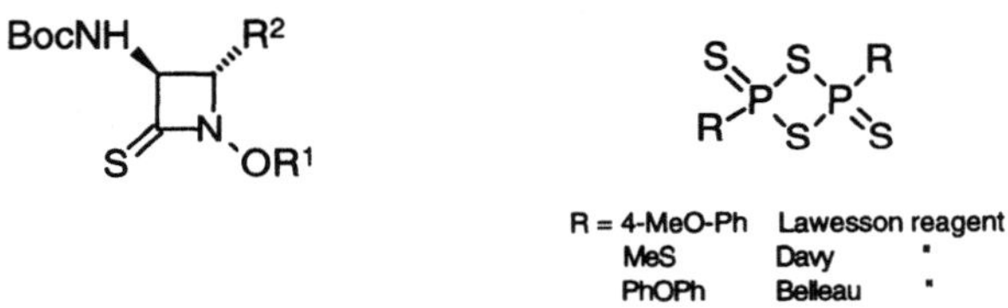

Scheme 2

A challenging thionation of the cyclic undecapeptide cyclosporin, a powerful immunosuppressant, has been attempted with Lawesson's reagent under a variety of conditions. The main product, out of four, resulted [30] from the replacement of amide oxygens by sulfur atoms located between residues 1 and 2. Chromatographic separation allowed, through NMR and crystal structure studies, the deduction of their conformations and correlation of them with their immunosuppresive activities. Later, higher yields were attained by sulfurising acetylcyclosporins [31].

The efficiency of a bis(disilyl)sulfide for thiolactam synthesis has been explored: prior activation of the amide into a Vilsmeier type intermediate was necessary [32].

Routes involving other sources of starting material have found success:

- Reaction of a carboxylic acid with an amine and *O,O*-diethyl dithiophosphoric acid [33]
- Addition of P_4S_{10} to nitriles in the presence of Na_2S [34]

- The high electrophilicity of dithioesters towards amines was applied to the synthesis of phosphonothioamides [35, 36] and thiohydroxamic acids [37]. Dithioacid salts may also be used as starting materials for this thioacylation with catalysis by boron trichloride [38]
- Addition of a carbanion [39] or an enolate [40, 41] to methyl isothiocyanate provided an entry to a new class of biologically active thioamides, bearing sulfinyl and pyridyl groups, introduced by Rhône-Poulenc-Rorer as cellular potassium channel openers

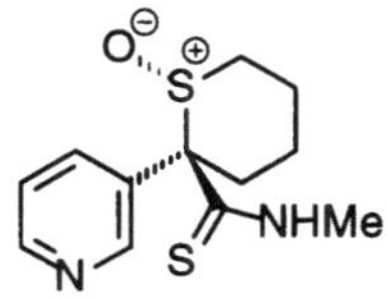

Scheme 3

Aprikalim

New routes have been designed by Katritzky and co-workers using benzotriazole derivatives. Alkylation of primary thioamides has been achieved on the nitrogen atom using an aldehyde as a source of the alkyl group [42]. A variety of thioamides is accessible by a one pot reaction of a Grignard reagent with carbon disulfide (in THF), followed by treatment with benzotriazole triflate and aminolysis of the activated thiocarbonyl intermediate [43, 44].

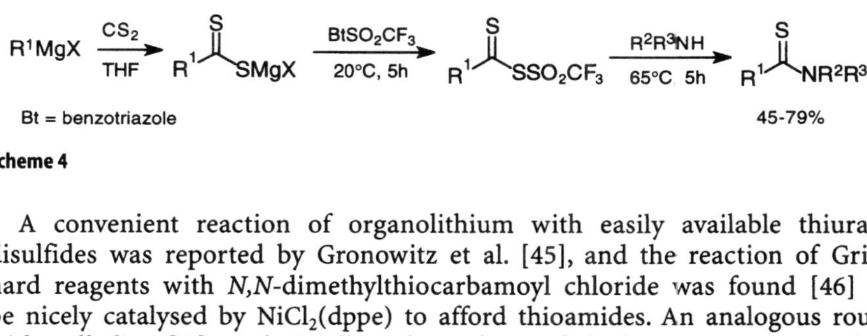

Scheme 4

A convenient reaction of organolithium with easily available thiuram disulfides was reported by Gronowitz et al. [45], and the reaction of Grignard reagents with *N,N*-dimethylthiocarbamoyl chloride was found [46] to be nicely catalysed by $NiCl_2$(dppe) to afford thioamides. An analogous route with palladium[II] catalysis allowed Hartke et al. [47] to prepare a number of interesting α-acetylenic thioamides.

Thiopeptides represent an active field of research which has recently been reviewed by Hoeg-Jensen [20]. Two approaches are considered: thionation of peptides, as reported above, and coupling reactions. The latter generally involves an α-aminodithioester and offers the perspective of automated synthesis, but it revealed some drawbacks, the main one being racemisation. Improved routes to α-aminodithioesters were sought. The reagent P_4S_{10}/NaF, presented above, proved convenient for their synthesis, with retention of configuration under strictly defined conditions [48]. The condensation of α-aminodithioesters with the alkali salts of α-aminoacids was much more rapid than with the corresponding esters [48]. Addition of DMAP (5 equiv) to a mixture of an α-aminodithioester and a peptide linked to a polymer led to coupling in good yields and with low epimerisation [49].

New thioacylating reagents were developed. A combination of monothiocarboxylic acids with a PyBOP type reagent, containing phosphorus with a potential P=O bond formation as the driving force [50, 51], works efficiently. A thionoacid derivative of nitrobenzotriazole provided a mild and racemisation free coupling with phenylalanine methyl ester [52]. This shows that this route is not a "dead end" as had been feared for some time [53].

2.2 Dithioesters

As classical routes (using CS_2, H_2S or LR) work efficiently [3, 54–56], the attention in this field has concentrated on the synthesis of dithioesters bearing complementary functional groups.

α-Amino dithioesters are accessible from the corresponding chiral nitriles and their reaction with thiols in acidic medium followed by sulfhydrolysis of the intermediate thioimidoesters [57]. The question of racemisation was addressed: the α-proton has a rather low pKa (9.05 for ethyl *N*-benzyloxycarbonyldithioalanin), and the extent of epimerisation was found to depend upon the nature of the R^2 group. Similarly, α-azidodithioesters were prepared for the first time [58] from nitriles. A stepwise route to α-amino dithioesters was achieved from α-amino amides, involving P_4S_{10} and H_2S sulfurisation in the presence of NaF [48].

A number of β-hydroxydithioesters have been prepared by the aldol reaction of pre-formed enethiolates. As enantiopure (*R*)- and (*S*)-3-hydroxypentanedithioates could not be obtained by baker's yeast reduction of 3-oxoalkanedithioates, another enzymatic route was used [59], involving lipase kinetic resolution of racemic 3-acetoxypentanedithioates. α-Oxodithioesters have been conveniently prepared from methyl ketones, through the reaction of a pyridinium salt with sulfur and alkylation, but this method appears efficient only for aromatic dithioesters [60]. The classical β-oxodithioesters are furnished [61] by a Claisen condensation of ketone enolates with trithiocarbonates, a method which can be preferable to the condensation of CS_2 and subsequent alkylation.

Substitution of the acetoxy group of an azetidinone by a dithiocarboxylate afforded intermediates for 2-substituted penems [62].

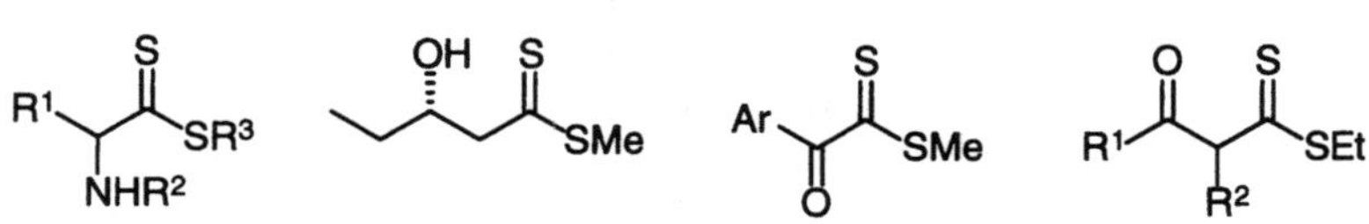

Scheme 5

Diconjugated dithioesters are still unknown: attempted syntheses led instead [63, 64] to 6-alkylthio-(2*H*)-thiapyrans, probably arising from intramolecular [4+2] cycloaddition of the expected dithioesters. In contrast, the

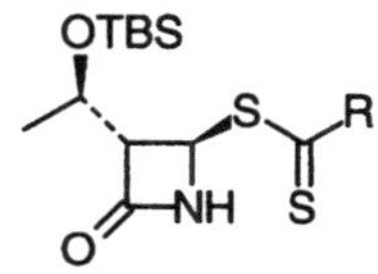

Scheme 6

monoconjugated methyl 2-propenedithioate has been prepared by FVT techniques and characterised by photoelectron spectroscopy [65]. Among the various conjugated thiocarbonyl compounds, the dithioester is the less thermally stable: it can be monitored only at low temperature or in the gas phase, whereas the thioamide is more stable.

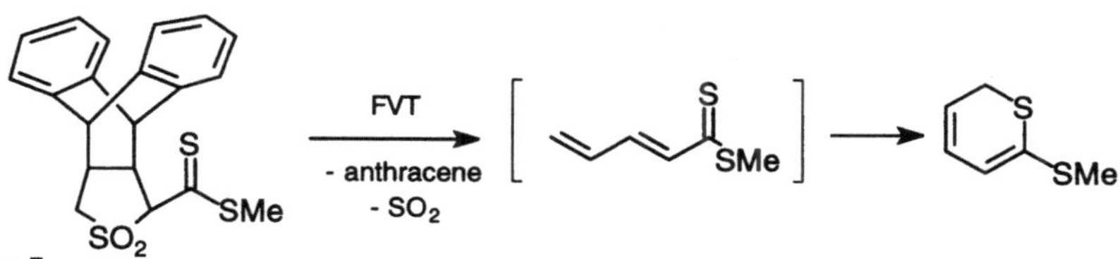

Scheme 7

2.3 Thionoesters

More work has been reported recently on this underexploited class of molecules. Vasella and his group [66, 67] have explored the synthesis of thionolactones in the sugar series. Glyconothio-*O*-lactones could be prepared by thermolysis of *S*-glycosyl thiosulfinates or by photolysis of *S*-phenacyl thioglucosides. The thionation of lactones with Lawesson's reagent was less effective, probably due to the poor tolerance of this reagent towards other oxygen atoms. A review article [68], in the context of brevetoxin B synthesis, describes the difficulties met for the thionation of dilactones bearing many oxygen groups.

Sulfurisation of a triolide from (*R*)-3-hydroxybutanoic acid with Lawesson's reagent (LR) in toluene led to the mono-, di- and trithioderivatives [69].

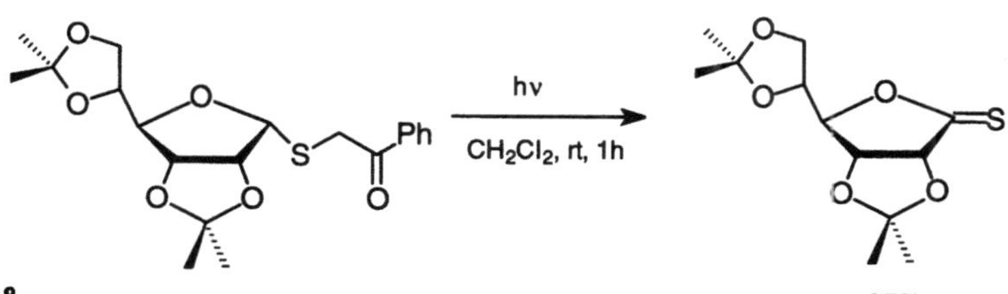

Scheme 8

Synthetic pathways which were used for dithioesters and reported above are also efficient for α-amino thionoesters [70] and β-oxothionoesters [61, 71, 72].

An expedient radical route has been reported by Bachi et al. [73]. Cyclisation of unsaturated xanthates initiated by n-Bu_3SnH/AIBN furnished thionolactones.

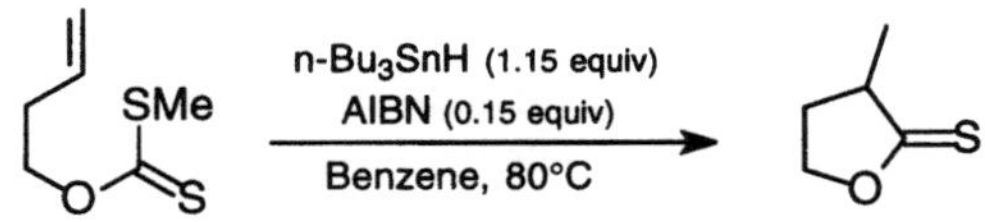

Scheme 9

2.4 Thioketones

An attractive reagent for the transformation of ketones into thioketones has been introduced by Degl'Innocenti and co-workers [74, 75]. The commercially available bis(trimethylsilyl)sulfide works efficiently at room temperature, with catalysis of $CF_3SO_3SiMe_3$ in acetonitrile, for the production of aromatic thiobenzophenones, adamantanethione and 2-cyclenethiones.

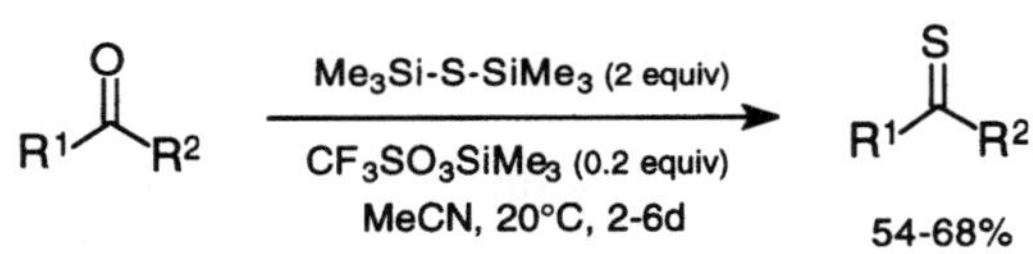

Scheme 10

Lawesson's reagent was used for the first preparation of carotenoid thiones [76] from canthaxanthin, rhodoxanthin and echinenone (carcinogenic benzene could very probably have been replaced as solvent by toluene, xylene or THF!).

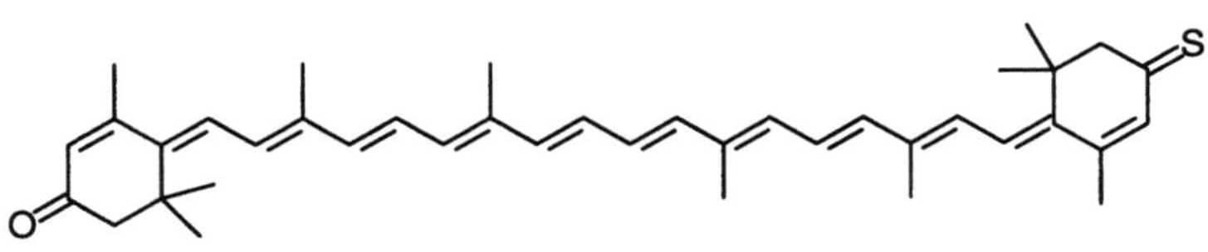

Scheme 11

The absence of a convenient method for the preparation of enethiolisable thioketones prompted us to fulfil this need. The development of aliphatic thioketones has so far been mostly restricted to examples with steric protection, such as thiocamphor and adamantanethione. Among the difficulties for a general synthesis of aliphatic and acyclic thioketones were: easy enethiolisation (enethiols are stable tautomers with thermodynamic stabilities close to those of the thione form), oligomerisation, and susceptibility to air oxidation. They were solved by the adaptation of a former German reaction using hydrogen sulfide in rather strong acidic medium. With a Lewis acid instead, zinc chloride, used in catalytic amounts, dimethylacetals of ketones were transformed into thioketones, devoid of tautomeric enethiols [77]. Flash chromatography furnished the expected thioketones, which could be kept in the cold for about a month, in reasonable yields.

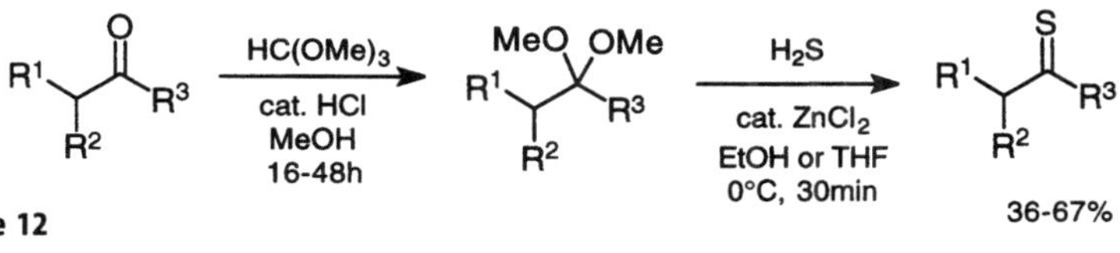

Scheme 12

This study was complemented by a selective synthesis of the tautomeric enethiols [77]. Aliphatic thioketones were deprotonated by LDA, silylated, and the resulting silyl vinyl sulfides were smoothly converted to enethiols by simple addition of methanol. These are *stable* compounds which do not equilibrate with thioketones, this behaviour probably related to the extremely mild conditions of the (easy) cleavage of the silicon-sulfur bond.

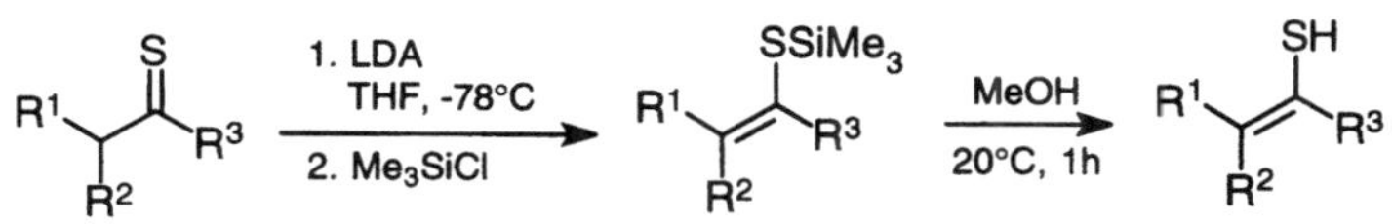

Scheme 13

Enethiols bearing three aryl groups were prepared by the reaction of a vinyl Grignard reagent with sulfur [78] or by sulfurisation of 2,2-diphenyl-1-arylethanone. The equilibrium between the thioketone and the enethiol is rapidly established, greatly in favour of the conjugated enethiol. This impressive difference with carbonyl compounds can be explained by bond energy differences. The first X-ray diffraction structure of an enethiol is disclosed in this study.

The sulfur analogues of acylsilanes have been prepared by treatment with H_2S or Lawesson's reagent [79–81].

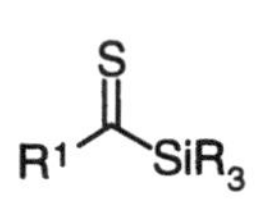

Scheme 14

Fascinating chemistry has been developed by Nakayama and Ishii on the first dithiiranes, which were prepared by oxidation of dithietanes and subsequent rearrangement. It was recently reviewed [82, 83]. Monodesulfurisation of dithiiranes was achieved with triphenylphosphine or triethylamine to yield the corresponding thioketones [84–86].

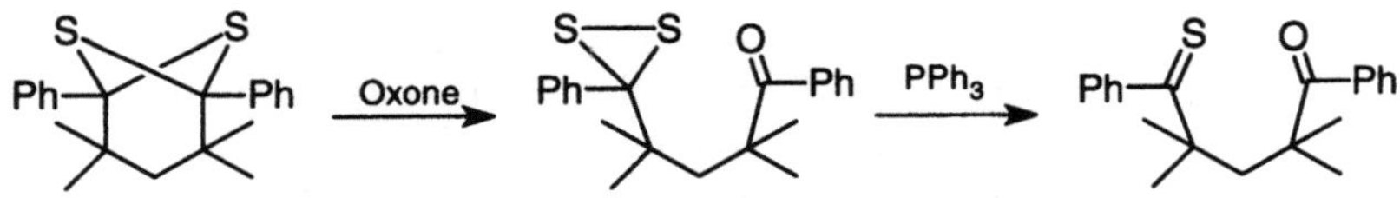

Scheme 15

Thioketones bearing an α carbonyl have been reported [87]. As these species are produced in situ and trapped by cycloaddition, this chemistry is reported in the corresponding section (Sect. 3.8). Sulfurisation of diketones by B_2S_3 led to dithioketones, which were later nicely found by the Steliou group [88, 89] to be capable of ejecting diatomic disulfur (S_2) via an elusive 1,2-dithietane.

Another route to this type of molecules was studied by Nakayama et al.: addition of elemental sulfur to alkynes with bulky substituents provided stable 1,2-dithietes [90]. The dithioxo forms were not monitored in these cases.

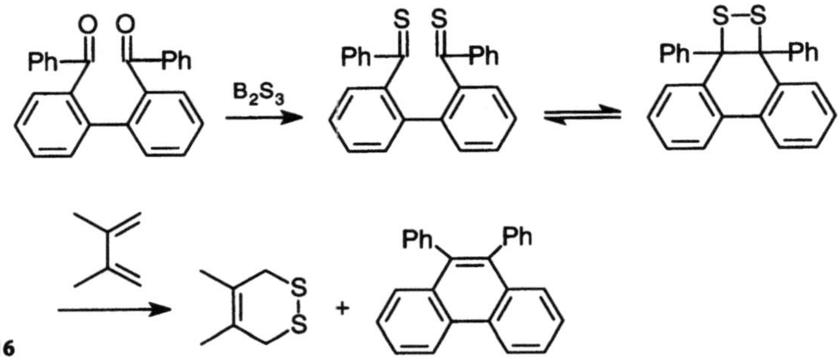

Scheme 16

2.5 Thioaldehydes

Among the compounds reported in this article, thioaldehydes are the least stable. They will not be reported in detail here as a recent series includes an up-to-date review by Okazaki [4].

A variety of examples has been prepared under conditions where a trapping agent, such as a diene, is present. This chemistry is described below in Sect. 3.8 on [4+2] cycloaddition reactions.

More information has recently appeared on the first stable aromatic thioaldehydes elegantly synthesised by the group of Tokitoh and Okazaki in Tokyo: molecular structures, detection and studies of rotational isomers.

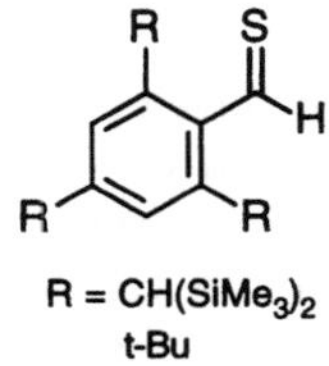

Scheme 17

2.6
Sulfines

Thiocarbonyl oxides are a subject of active investigation. The natural occurrence of sulfines and related compounds in plants of the genus *Allium* (onion, garlic, etc.) is included in a superb and extensive review by Block [91]. Two detailed papers [92, 93] report the isolation of zwiebelanes from onions and their chemical synthesis involving intermediate sulfines produced by oxidation of di-1-propenyl disulfide, subsequent sulfoxide accelerated [3.3] sigmatropic shift and the [2+2] cycloaddition of the C=S and C=S=O moieties. A further article [94] provides a great deal of information on the mechanism of formation of *(Z)*-propanethial *S*-oxide, the lachrymatory factor of the onion, as well as its chemical synthesis and reactions. Techniques of analysis of the volatiles of onions have been further improved [95].

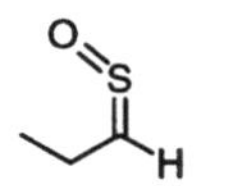

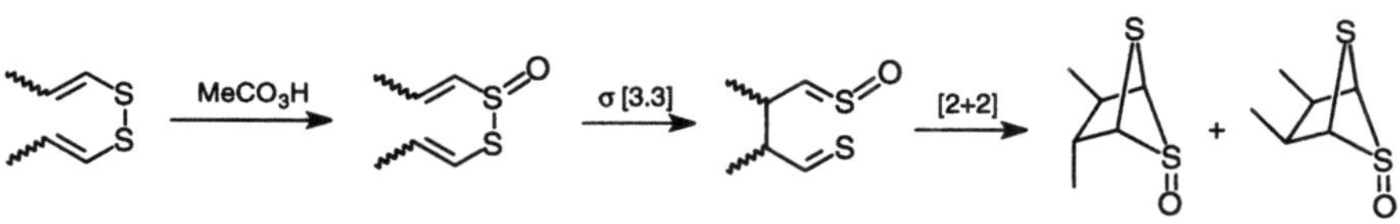

Scheme 18

This facile type of sigmatropic of rearrangement appears general. Julia and his group [96–98] have prepared and characterised a number of unsaturated sulfines. Instead of an oxidation route to the starting sulfoxides, they performed the reaction of vinyl Grignard reagents with allyl sulfenates at −78 °C. In most cases; the subsequent [3.3] shift is so rapid that the intermediate unsaturated sulfoxides were not even evidenced: sulfines were isolated instead.

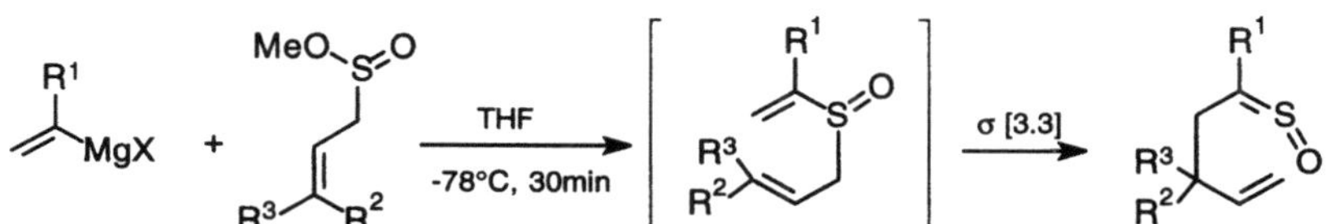

Scheme 19

Examples with a variety of frameworks have been reported starting from allenic sulfenates [98] and 1,3-dienic sulfenates [96].

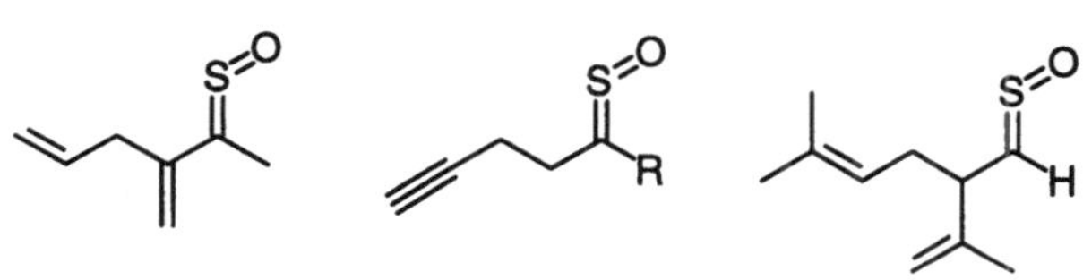

Scheme 20

A novel synthesis of α-unsaturated sulfines has been introduced by Braverman et al. [99]. Et_3N or DABCO treatment of allylic and benzylic trichloromethyl sulfoxides triggered the elimination of chloroform and formation of the sulfines. It must be stressed that these sulfines are thermally relatively stable, and this stands in high contrast to the corresponding thiocarbonyl compounds: unsaturated thioaldehydes cannot be monitored under the same experimental conditions and have to be used at very low temperature or trapped in situ. The first synthesis of thioacrolein *S*-oxide was achieved by flash vacuum thermolysis of an anthracene allyl sulfoxide [100], and both isomers in a (*Z*:*E*) ratio of 78:22 were characterised by NMR spectroscopy at −60 °C.

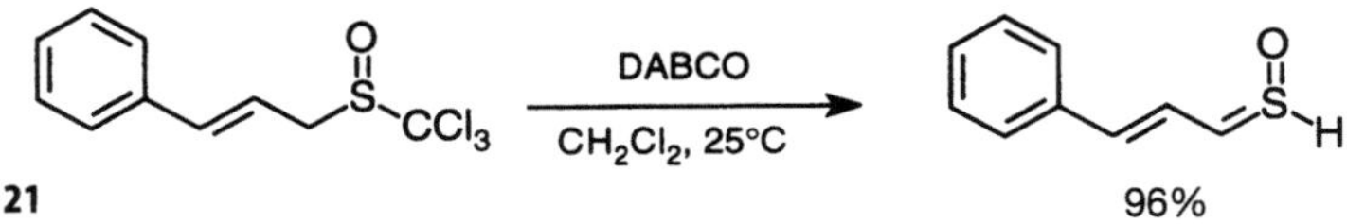

Scheme 21

A third method of access to sulfines is the oxidation of thiocarbonyl compounds. When the starting material is available it is an attractive route. There has been some dispute in the past whether enethiolisable thiocarbonyl derivatives would lead to the corresponding sulfines or to divinyl disulfides [101, 102]. It is now clear from our research that, even if the C=S molecules bear highly acidic α-protons, oxidation occurs on C=S and does not touch the α-protons. There are many examples of this behaviour. The most easily enethiolised compounds are thioketones. We have shown that their reaction with a peroxycarboxylic acid, mCPBA, is very fast at 0 °C and quantitatively provides the corresponding sulfines [103]. In many examples the aliphatic sulfines are not very stable and have to be used in subsequent reactions that will be faster than their decomposition ($t_{1/2}$ from some hours to days).

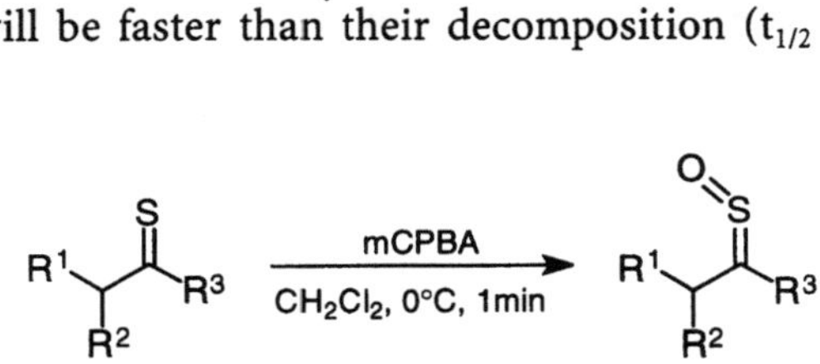

Scheme 22

A wide variety of sulfines has been produced by the oxidation reaction with mCPBA starting from dithioesters [104, 105], trithioperesters [106], heterocyclic C=S compounds [107], thionoesters [108] (see also [109]), dithiocarbonates [110] and trithiocarbonates [111]).

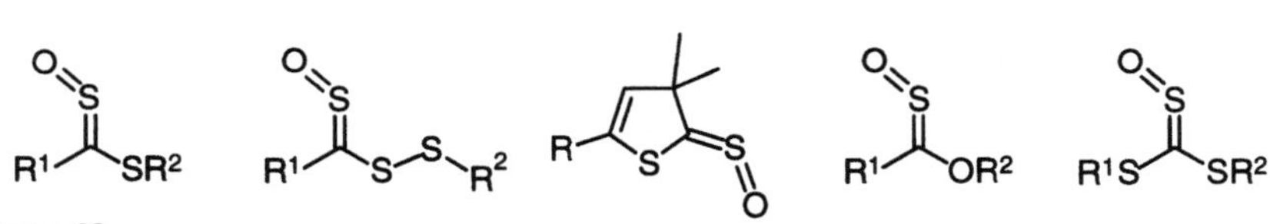

Scheme 23

Zwanenburg and his group [105] have addressed the question of the tautomeric interconversion of aliphatic sulfines into vinyl sulfenic acids. They have synthesised dithioesters bearing an α asymmetric carbon centre. Oxidation with mCPBA did produce chiral sulfines, which did not racemise except when a phenyl group was attached to the α carbon.

The dithiete isomers of bis(thioketones) have been oxidised [112] to afford the first α-bis(sulfines), fully characterised by X-ray analysis. The major isomers have an *(E,E)* structure. For the monoxidation, a dithiete monoxide was evidenced and shown to equilibrate with the α-thioxo sulfine. These results confirm that oxidation of the C=S moiety stabilises this weak 2p-p double bond.

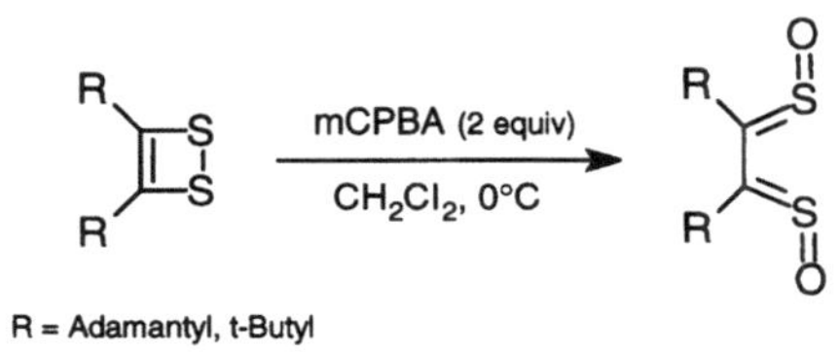

Scheme 24

Machiguchi and co-workers [113] have succeeded in the first synthesis of tropothione *S*-oxide, which is amazingly stable. Physical data led the authors to propose a charge reversion of the π electron distribution on the carbon as compared to the parent sulfine $H_2C=S=O$.

Interesting structures were obtained from cycloalkyl silyl thioketones which yield the corresponding (*E*)-sulfines. The latter can be desilylated to sulfines bearing a hydrogen, thus formally deriving from thioaldehydes, a route that is not directly feasible due to the instability of thioaldehydes under these conditions. However, starting from the rare examples of stable thioaldehydes, which bear aromatic groups with very bulky substituents, mCPBA oxidation led to corresponding *(E)*-sulfines [114].

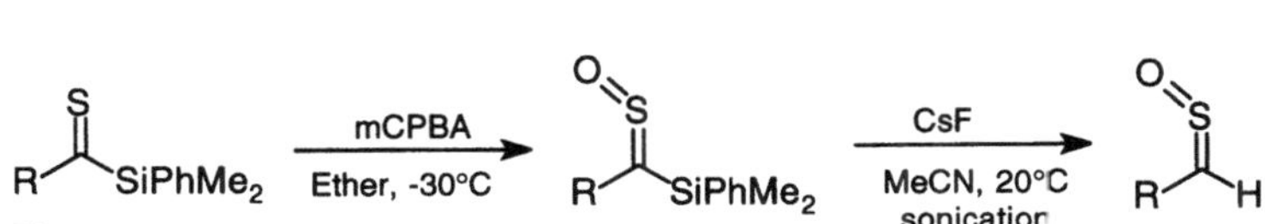

Scheme 25

Pentafluorophenyl thioketones are available [115] from the reaction of the corresponding fluoroketones with B_2S_3 formed in situ. Oxidation with monoperphthalic acid generated the sulfine, the X-ray structure of which revealed a dimer in the crystal lattice.

A number of studies dealing with α-oxo sulfines has appeared. The stabilities of the attractive molecules usually do not allow isolation. The strategy used for the manipulation of unstable thioaldehydes was applied here. A precursor is submitted to an elimination reaction (or a cycloreversion) to generate the sulfine in the presence of a diene, and it is the product of

[4+2] cycloaddition which is isolated. This chemistry is treated in Sect. 3.8.3.

Finally I would like to mention the studies in search of the analogous thiocarbonyl *S*-sulfides. This field, which was pioneered by Huisgen and Senning, is currently also under investigation by three Japanese groups. The structure itself has been directly detected by spectroscopy but various reactions led to these elusive intermediates which were trapped by a carbonyl group [84], a dipolarophile [116] and a dienophile [117]. A full report on the chemistry of these *S*-sulfides developed by Huisgen and Rapp [118] provides a great deal of information and raises many new questions.

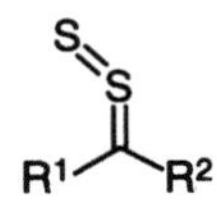

Scheme 26

3
Chemistry and Applications

The main feature of thiocarbonyl compounds is their high reactivity. For instance their strong electrophilicity, relative to carbonyl compounds, is related to their low lying LUMO which causes an important reduction of the gap with occupied orbitals of nucleophiles [119]. At the same time, they are more nucleophilic in relation to their high HOMO. A number of examples reported below illustrate this.

3.1
Enethiolates and Their Reaction with Electrophiles

The sulfur analogues of enolates have recently received attention in the context of synthetic applications. Thiocarbonyl compounds bear α-protons which are rather acidic. Kresge et al. [120] has shown that their pKas are 10 units less than those of carbonyl compounds. Thus enethiolates are easily formed with a variety of bases, and they exhibit thermal stability [1]. They are ambident nucleophiles and the sulfur vs carbon regiochemistry has been rationalised by Anh [119] using frontier orbital treatment.

The main class of electrophiles which reacts on the sulfur atom of enethiolates are alkyl halides. This was applied by Vallée and Tchertchian [121] in a one pot synthesis of hydroxy-ketenedithioacetals which elegantly uses the preceding features. Deprotonation of alkyl dithioacetates by LDA at −78 °C provided enethiolates which were treated by aldehydes. Comparable to the case of enolates, the aldol reaction takes place on the carbon atom. When water is added to the reaction mixture, 3-hydroxyalkanedithioates are obtained. However, if the quench is replaced by an alkyl halide addition, hydroxyketenedithioacetals are obtained. Formation of these compounds arises from the intermolecular deprotonation of intermediate dithioesters (lower pKa) by the lithium aldolate moiety and subsequent formation of an enethiolate. Addition of an alkyl halide shifts the equilibrium in favour of ketenedithioacetals due to the higher nucleophilicity of the enethiolate as compared to the alcoholate. It is also remarkable that the proton transfer occurs with a high selectivity in favour of the "*cis*" enethiolate.

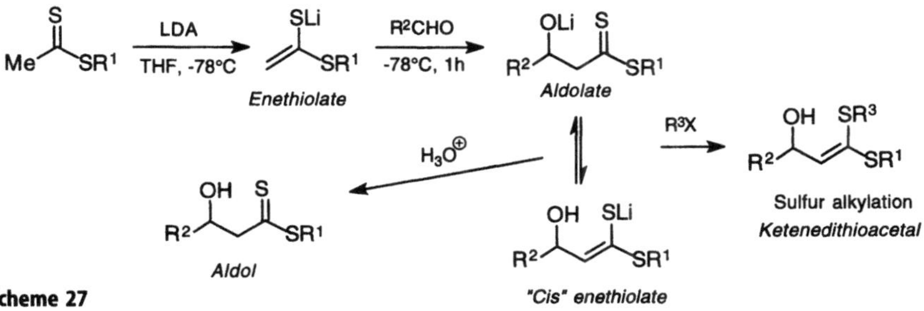

Scheme 27

An arylation reaction of potassium thioamide enethiolates was achieved in good yield by treatment with aryl iodides in the presence of $FeBr_2$ or with photochemical irradiation, perhaps through an $S_{RN}1$ mechanism [122].

The aldol reaction of dithioester enethiolates has been used for the synthesis of dithiolactones [123], chiral substrates for the Claisen rearrangement [124, 125], and oxathianes [126]. Thioamide enethiolates may be employed in this reaction as well as shown with β-thiolactams [26], or with a precursor of the antitumour agent vinblastine [127].

Enethiolates are soft nucleophiles which react in a conjugate fashion with Michael acceptors [1, 128, 129]. Thioamide enethiolates are a borderline case

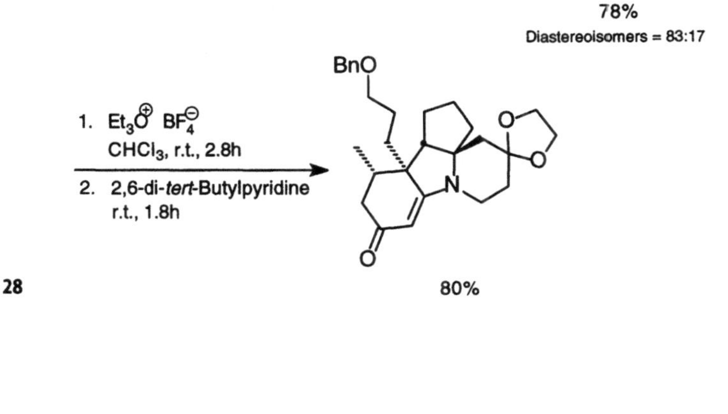

Scheme 28

cf 1,4 vs 1,2-addition [130, 131]. Further studies on the behaviour of titanium derivatives have been reported [132]. The conjugate addition was used efficiently in a total synthesis of a *Daphniphyllum* alkaloid [133] and of a phenanthridin-3-one derivative [134] of therapeutic interest.

A secondary thioamide has been reacted with methyl acrylate in the presence of sodium hydroxyde to afford a 1,4-addition product on the nitrogen atom [135].

Amination of enethiolates has only been recently tackled by Beslin and Marion [136] and Hartke et al. [137]. Silylketene dithioacetals or lithium enethiolates react with azodicarboxylates to yield α-hydrazino-dithioesters, thionoesters or thioamides. Homologous dithioesters, bearing an amino moiety on the β position, have been prepared from the reaction of silylketene dithioacetals with an imine in the presence of a Lewis acid [138].

The reaction of non-enethiolisable dithiobenzoic esters with LDA led to reductive dimerisation [139]. In the presence of methyl iodide, 1,2-bis(methylthio)stilbenes are formed, probably through single electron transfer from LDA, formation of a radical anion, dimerisation to a dithiolate and alkylation.

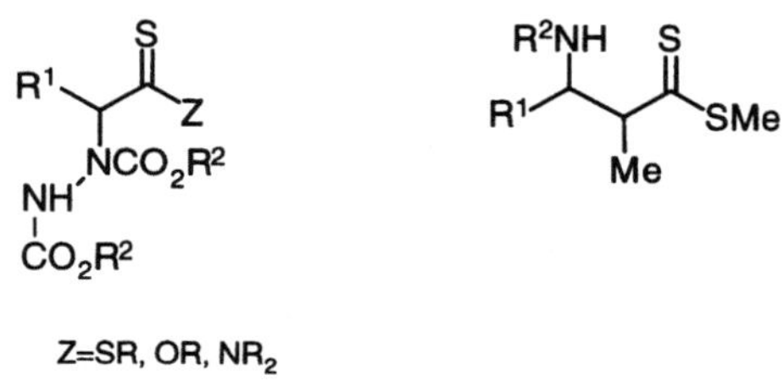

Scheme 29

The ease of enolisation of sulfines and the use of enesulfenates are a debated topic. A contribution is the report that vinylsulfenic acids, tautomers of enethiolisable sulfines, have been trapped intramolecularly by a double or triple carbon-carbon bond [140], as shown in the accompanying example.

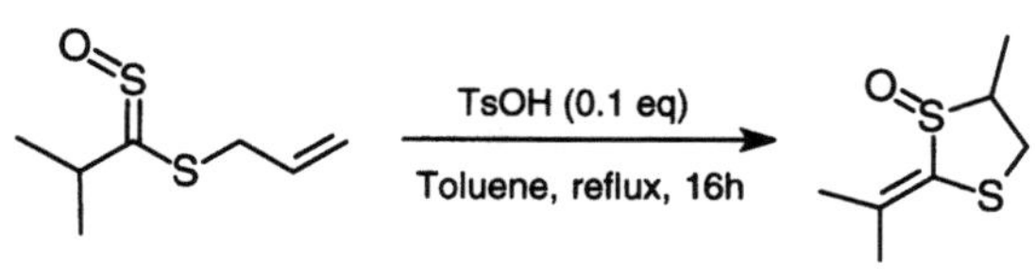

Scheme 30

In the next section the formation of acyl anion equivalents by nucleophilic addition to thiocarbonyl compounds is discussed. A direct and non-classical route to thiocarbonyl anions has been achieved [141]. Treatment of a thiocarbamoyl chloride by lithium powder, in the presence of both naphthalene and the carbonyl compound to which the intermediate will be added, led to α-hydroxythioamides.

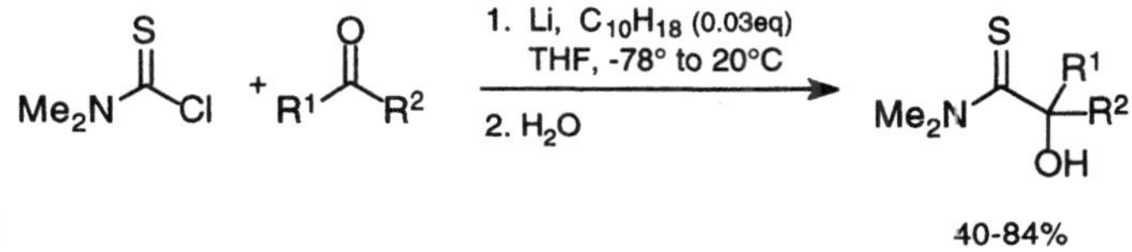

Scheme 31

The same intermediate was also formed by direct deprotonation of *N,N*-dimethylthioformamide with LDA, and was used to prepare α-oxo thioamides by acylation with carboxylic esters [141]. Deprotonation of the formed thioamides laterally takes place on the *N*-methyl group and intramolecular addition of the anion to the carbonyl group provides a new entry to β-thiolactam rings.

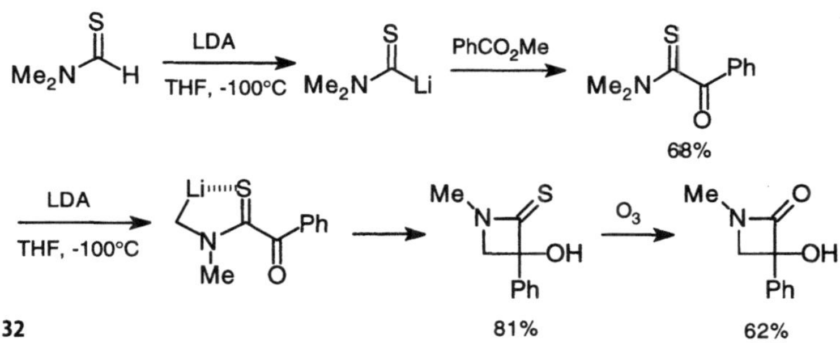

Scheme 32

3.2 Addition of Carbanions: Thiophilic vs Carbophilic Reactions

The thiophilic addition of carbon nucleophiles to thiocarbonyl compounds was a topic of intense investigation during the 1980s.

The high reactivity towards nucleophiles, as compared to carbonyl compounds, is easily explained by frontier orbital treatment [119]. However the regioselectivity of addition, whether on sulfur or on carbon atoms, has not yet been fully rationalised. However this has not prevented numerous successful applications [1].

A pioneer in this field, Peter Beak, has tested [142] the intramolecular addition of an aryllithium bearing a dithioester group on a chain of variable length, in relation to geometrical features, and these processes were compared with the thiophilic radical addition of *n*-Bu_3SnH in the presence of AIBN. Thus, ethyl iodoaryldithioalkanoates were treated with 2.2 equivalents of *t*-BuLi. Predominant halogen-metal exchange took place with subsequent thiophilic addition and cyclisation.

Further examples of thiophilic addition of methyllithium and carbophilic addition of allyl- or benzyllithium and allyl Grignard reagents with 1,3-thiazole-5(4*H*)-thiones were reported [143]. Perfluorodithioesters were reacted with alkylmagnesium bromides, providing a new entry to perfluoroketene

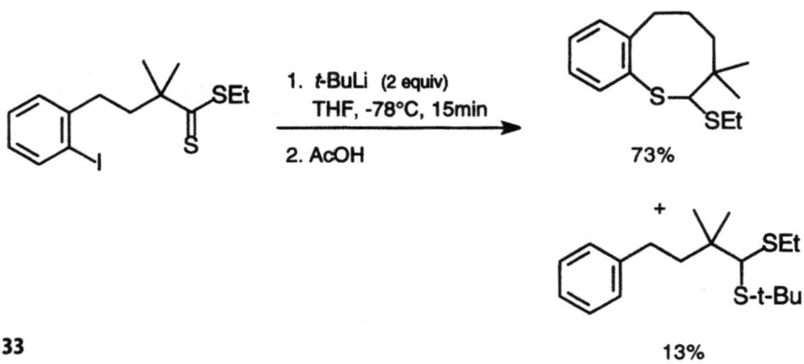

Scheme 33

dithioacetals, through a thiophilic addition and subsequent elimination of an α-fluorine atom [144].

A new class of nucleophiles have been introduced for sulfur addition. Degl'Innocenti and his group [145, 146] have shown that allyl or benzylsilanes, in the presence of tetra-*n*-butylammonium fluoride, reacted in a thiophilic fashion and led to allyl sulfides or dithioacetals. It is remarkable that this selective reaction is general for a large variety of thiocarbonyl compounds: thioketones [145], dithioesters [146], and even with the normally sluggish trithiocarbonates [145]. With substituted allyl silanes retention of configuration of the allyl chain is observed. It is noteworthy that the possible [2,3] sigmatropic shift of the intermediate anionic species was not observed.

Scheme 34

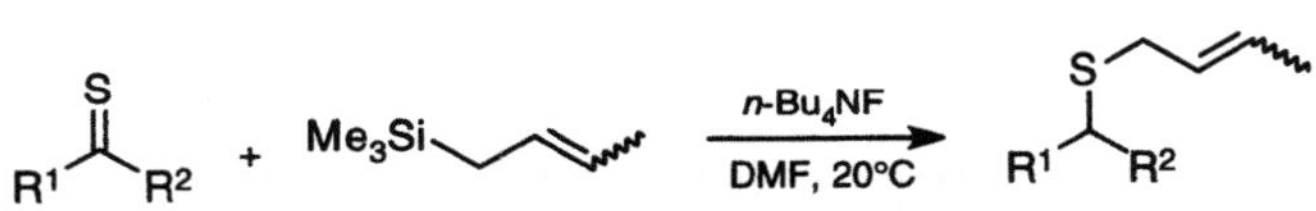

Thionoesters are a class of compounds which almost always react by carbophilic addition with organolithiums. This was confirmed [147] with thionolactones and 1-propenyllithium, but some cases of enolisation have also been observed. A newly prepared glyconothio-*O*-lactone was submitted to the action of methyllithium [66]. Two products were formed, arising from carbon addition and enolisation. Lithium dimethylcuprate led to a chemoselective addition with a moderate diastereoselectivity.

Thioamides are usually sluggish towards organolithiums or Grignard reagents. It has been recently shown that primary thiobenzamides react with methylcerium dichloride to afford good yields of tertiary amines [148].

A general way to achieve thiophilic addition is to use sulfines [149]. The sulfur atom of thiocarbonyl oxides is selectively attacked by carbanions. We have applied this property to aliphatic dithioester oxides and shown [150] that, despite the probably high acidity of the α-protons, thiophilic addition of alkyllithiums is observed, leading to stabilised carbanions. After treatment with electrophiles, dithioacetal oxides were obtained. From a synthetic point of view, it is worthy of note that these adducts are much more easily converted to aldehydes or ketones than are their dithiane analogues. Simple treatment with a mineral acid, such as HBF_4, or even mere standing of the compound at ambient temperature, is sufficient for unmasking the latent functionality. Thus sulfines are a starting material for the acyl anion or "*Umpolung*" chemistry, with the main difference that the intermediates arise here from an addition reaction instead of a deprotonation. This brings new prospects for chemoselectivity.

A selective reaction was also observed in the heterocyclic series with 1,3-thiazole-5-(4*H*)-thione oxides and alkyllithiums [107]. A more sluggish reaction was observed with Grignard reagents.

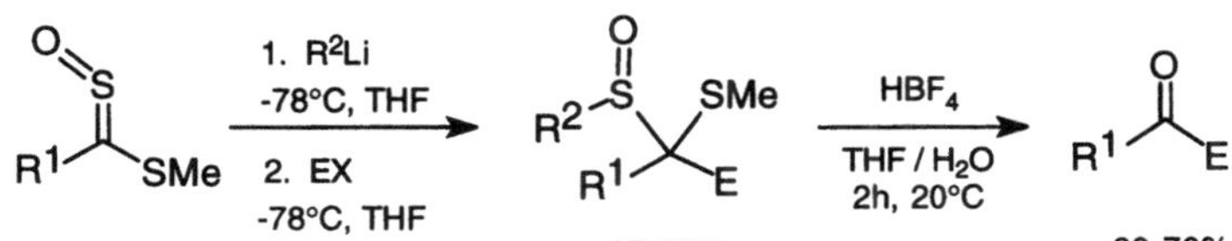

Scheme 35

Trithiocarbonate *S*-oxides are reactive towards alkyllithiums [111]: thiophilic additions were carried out at –78 °C. The resulting carbanions, stabilised by three sulfur groups, were quenched by water or by other electrophiles to afford trithioorthoester oxides. With enones, 1,4-addition was observed: elimination of an alkanesulfenic acid led to *β*-oxoketenedithioacetals which could be transformed into 4-oxoalkanethioates. This "*Umpolung*" route allows the formal use of an (alkylthio)carbonyl anion.

Allyl silanes may also be employed for thiophilic addition to various sulfines, obtained by oxidation of thioketones, dithioesters [151] and trithiocarbonates [111]. It provides a new entry to allyl sulfoxides.

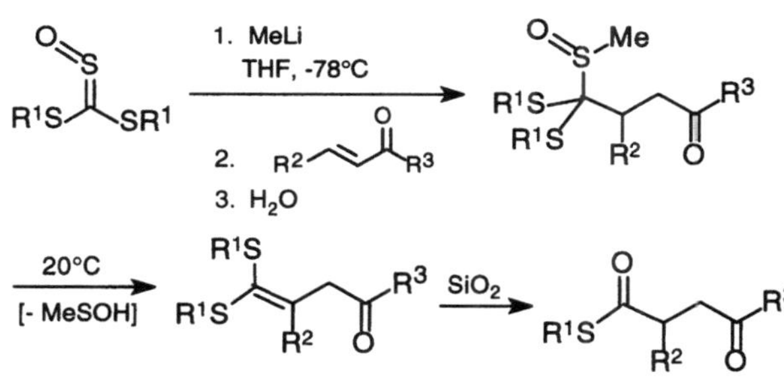

Scheme 36

3.3
Reactions of Heteroatomic and Other Nucleophiles

A variety of nucleophiles have been reacted with thiocarbonyl compounds, usually for synthetic purposes exploiting the enhanced reactivity of the 2p-3p double bond between carbon and sulfur atoms. In contrast to the preceding part, most nucleophilic additions reported took place on the carbon atom: very few cases of thiophilic addition were evidenced.

The reaction of a thioxosteroid with hydrazine hydrate is much faster on C=S than on C=O. In the absence of a catalyst it takes place in a matter of minutes to provide the corresponding monohydrazone [152]. The rare stable aromatic thioaldehydes are also quite reactive [114].

As mentioned above for the preparation of thioamides, the reaction of dithioesters with amines is generally rapid and efficient. Kinetics have shown [153] that this reaction involves two molecules of amine and consists of the following events: nucleophilic addition, amine assisted prototropy and decomposition of the neutral tetrahedral intermediate. This study was undertaken with a view to polythioamide synthesis.

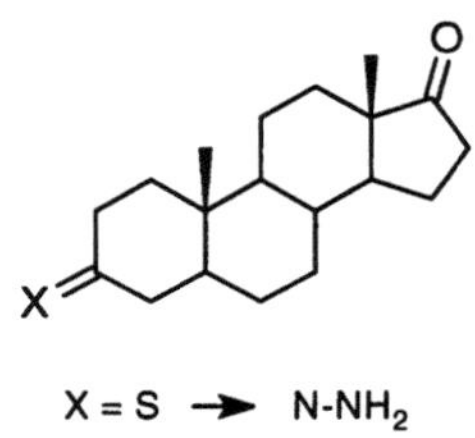

Scheme 37

In contrast, the reaction of thionoesters bearing a β-carbonyl group with primary amines does not lead to thioamides, but instead to α-oxo aminoketene acetals, arising from elimination of H_2S rather than ROH [71].

Reaction of thiolactams, in the sugar series, with amines allows the preparation of amidines which were desired for evaluation as glycosidase inhibitors [154].

The behaviour of phosphonodithioformates has been extensively examined by Masson and recently reviewed [36, 155]. Among the nucleophiles used, phosphites have been shown to provide new phosphonium ylides. Other reagents (organometallics, radicals, amines, thiols) have been employed for carbon-carbon bond formation and synthesis of new functionalised phosphonates, of potential biological interest.

The reduction of the dithioester group to a thiomethyl group has been achieved by sodium borohydride in acetonitrile at reflux [156, 157], offering an easy synthesis of thiomethyl phosphonates. Milder conditions or use of BMS provided the hemithioacetals.

The mechanism of the Wittig reaction of phosphoranes with carbonyl compounds has fascinated chemists and remains a matter of discussion. Though the sulfur equivalent is very much less used, this reaction is known

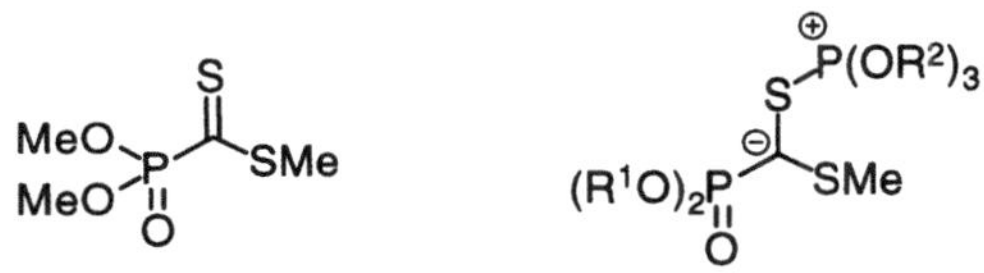

Scheme 38

to proceed, in the case of enethiolisable thioketones, to lead mainly to thiiranes and to triphenylphosphine. In a very recent mechanistic study, Erker and his group [158] were able to demonstrate an intermediate thiaphosphetane. Depending on the temperature, this intermediate decomposed either to a thiirane and a phosphine or to the Wittig alkene and a phosphine sulfide.

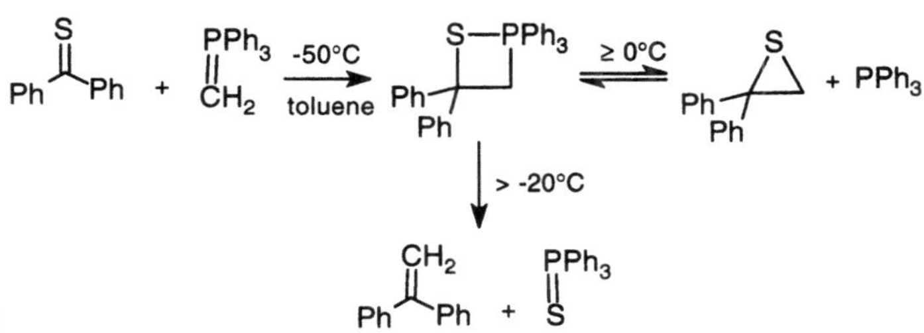

Scheme 39

The enzymatic reduction of a thiocarbonyl compound has been investigated [159] for the first time, in order to provide a new route for enantiopure thiols, molecules which are currently needed for asymmetric synthesis. Reaction of easily available β-thioxoesters with baker's yeast under classical conditions did furnish the expected thiols, but with lower enantiomeric purity and moderate conversion rate, due to the competitive "hydrolysis" of the thioxo group into a carbonyl leading to an alcohol. However, conditions (ethyl acrylate, dry yeast) were found to improve the production of (*S*)-ethyl 3-mercaptobutanoate. Cyclic thioxo esters led to high stereoselectivity of *cis* (1*S*,2*S*) products, but with moderate chemical yields.

Exceptions to the carbophilic addition of heteronucleophiles to thiocarbonyl compounds were demonstrated with thiols and thioaldehydes [160] or dithioesters [36] bearing an α-electron withdrawing group, which undergo selective thiophilic addition.

Reversal of reactivity was also sought with sulfines, in which the sulfur atom is expected to be strongly electrophilic [161, 162]. Results with amines and alcohols were recently disclosed. The sulfine of thiocamphor reacts with primary amines at room temperature, despite the steric hindrance, to give the corresponding imines [161]. A similar observation was made with thioaldehyde *S*-oxides [163]. The carbophilic addition was also proposed to explain the products of the reaction of amines with a sulfine bearing an ester group on the a position [162], in the context of the inhibition activity of a sulfinamoyl ester towards cinnamoyl alcohol dehydrogenase. The role of a hydrogen bond between the amine and the S-O bond to favour

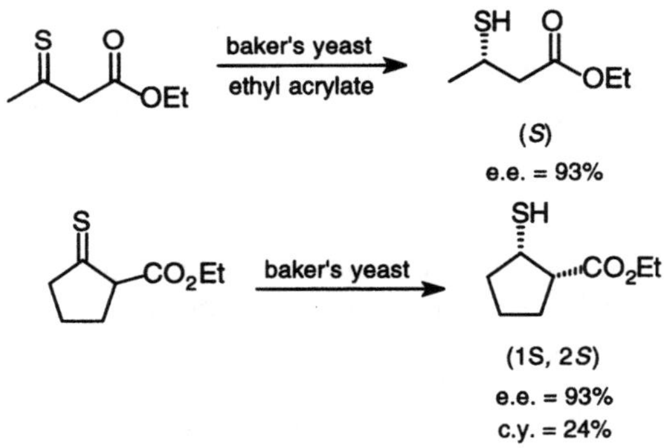

Scheme 40

this sense of addition was considered. No exception to the general trend was observed with dithioester oxides which provided thioamides after carbophilic attack of amines and elimination of sulfenic acid [161].

The reaction of alkoxides with thioaldehyde *S*-oxides, generated from sulfinates, was carefully examined by Baudin et al. [163]. A number of products were isolated, formation of which was explained by carbophilic addition of the alkoxide anions.

3.4 Some Miscellaneous Reactions

The reaction of a lanthanide metal, ytterbium, has been explored by the group of Fujiwara [164–166]. Among the various processes (coupling, desulfurisation, etc.) which were evidenced, a nice application was achieved by successive reaction of thiobenzophenone with ytterbium, and treatment of the intermediate with a dihalide. Thiolanes, for instance, were thus obtained in an expedient manner.

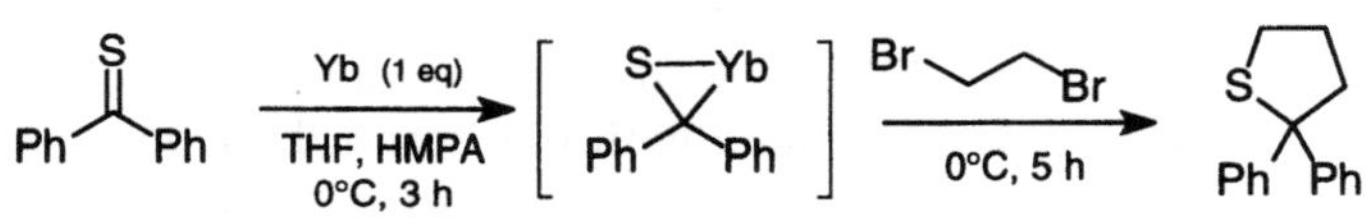

Scheme 41

The formation of thiocarbonyl ylids by the reaction of metallocarbenoids with thiocarbonyl compounds has not been extensively studied, as noted by Padwa and Weingarten in a review [167]. However, formation of rings by interaction of these compounds has recently received some attention. A key step in the synthesis of a polyhydroxylated indolizidine alkaloid related to castanospermine is the reaction of diazoketone with a thioamide, in the presence of rhodium acetate [135]. After desulfurisation an enaminone was obtained.

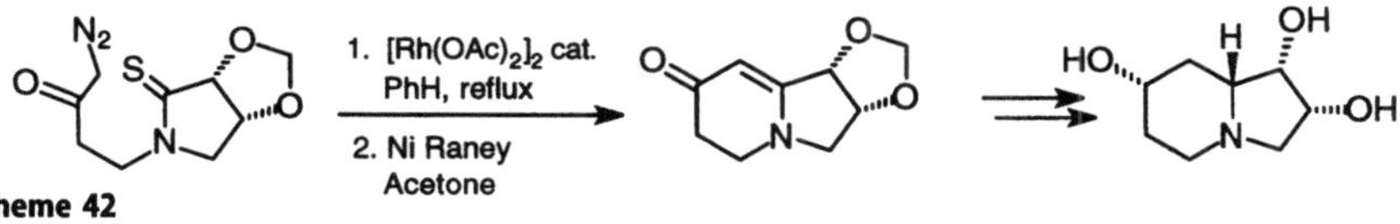

Scheme 42

A recent synthesis of (±)-supinidine involved the same reaction as a key step to form a 5-membered ring [168]. An intermolecular reaction of a carbenoid with a thionolactone in the *manno*-series led to a thiirane and the alkene product of desulfurisation [67]. A similar process was observed in the case of a dithioester leading, after desulfurisation, to a thiophene, and a reaction with a thioamide furnished a thiocarbonyl ylid which could be trapped by an electron poor olefin [169].

The reductive Raney nickel desulfurisation of thioamides is used routinely in the alkaloid series. It proved efficient for the synthesis of *Vinca* alkaloid related compounds [21], a vinblastine precursor [127], 19-hydroxytubotaiwine [170], and (–)-monomorine [171].

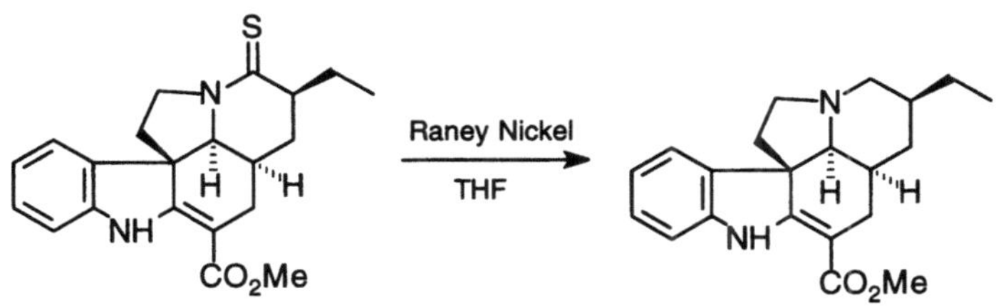

Scheme 43

3.5
Reactions with Electrophilic Reagents

Thioamides exhibit a pronounced nucleophilic character due to their nitrogen atom. Alkyl halides often react at room temperature to give thioimidoester salts which may be subsequently used. This reaction was applied [134] to the preparation of an enaminoketone, the cyclohexenone ring being formed by base treatment of a ketone and intramolecular reaction with the iminium salt. A similar sequence [135] was mentioned in Scheme 28.

The alkylation of thioamides is currently used for the Eschenmoser reaction which is discussed later in this article.

The increasing utilisation of thiocarbonyl compounds brings about the necessity for efficient procedures for their transformation into carbonyl compounds. A straightforward method [172] involves the treatment of a thioketone with 4-nitrobenzaldehyde in the presence of TMSOTf. Thus thiobenzophenones and thiocamphor are converted into the corresponding ketones, but aliphatic thioketones or thioamides are resistant under these conditions. In order to convert an easily accessible bis(thiocamphor), de-

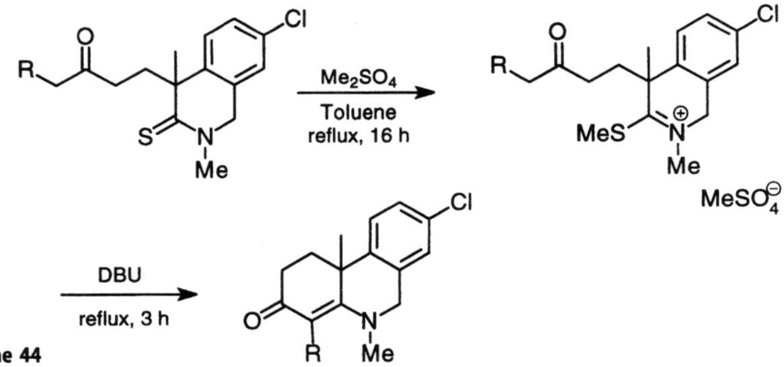

Scheme 44

sired for a new chiral ligand synthesis, Le Corre and his group [173] developed an alternative procedure. Conversion to the bis(sulfine) with mCPBA, and further oxidation with the $RuCl_3/NaIO_4$ combination provided bis(camphor) in a remarkable 93% yield.

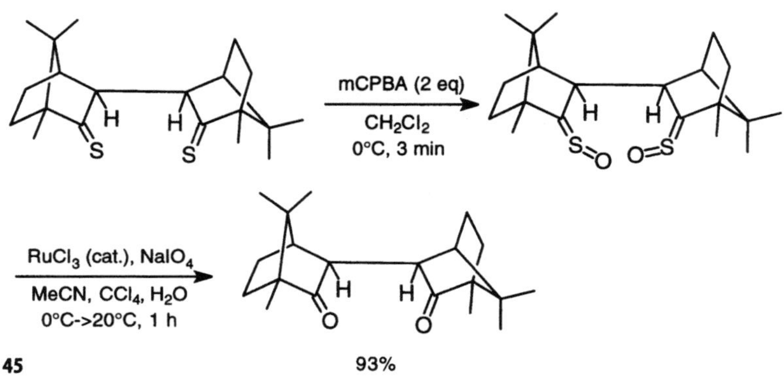

Scheme 45

The need for a variety of *organofluorine* molecules has encountered difficulties with oxygenated substitutes such as carboxylic esters due to their poor nucleophilicity towards electrophilic fluorinating reagents. This has led several groups to explore the conversion of the thiocarbonyl group into the CF_2 moiety. It was successful with thioamides [174], dithioesters [175–177] and thionoesters [178, 179], providing new routes to difluoro-sulfides, amines, ethers and alkenes.

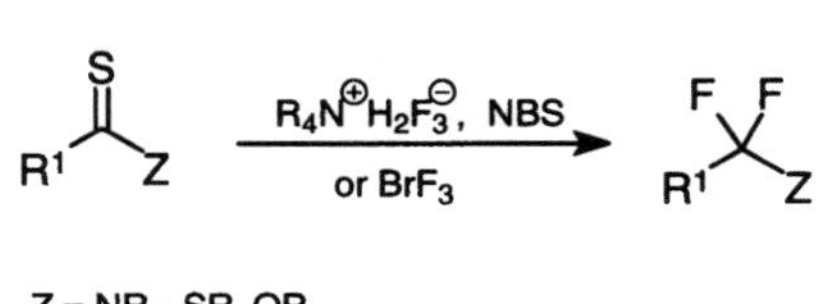

Scheme 46

The *Eschenmoser reaction* is extremely useful for the conversion of amides into enaminoesters via the thioamide reaction with α-haloesters, and triphenylphosphine mediated sulfide contraction, and we are fortunate that Shiosaki has published a thorough review on this topic [180]. The accompanying scheme shows a typical example for which an organometallic route with a lithium or a zinc enolate was not successful [181].

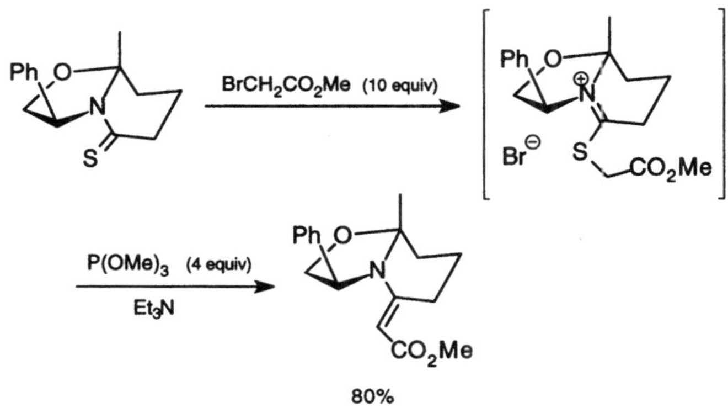

Scheme 47

Lhommet and his group [182] have tackled the difficult creation of tetrasubstituted C-C double bonds through the Eschenmoser reaction. They found conditions to overcome the unfavourable alkylation with secondary α-bromo esters: slow addition of triethylamine and triphenylphosphine to a solution of thiolactam and α-bromo ester. In this way the thioiminium salt was trapped as soon as it was formed.

Other reports also demonstrate the efficiency of this reaction for the synthesis of pyrrolidine prostacyclin analogues [183], (±)-indolizidine 209B [184], pyridinones [185] and pyrrolidine-2-ylidene esters [186], though in one case the structure of the substrate had to be modified to avoid an undesired reaction [187].

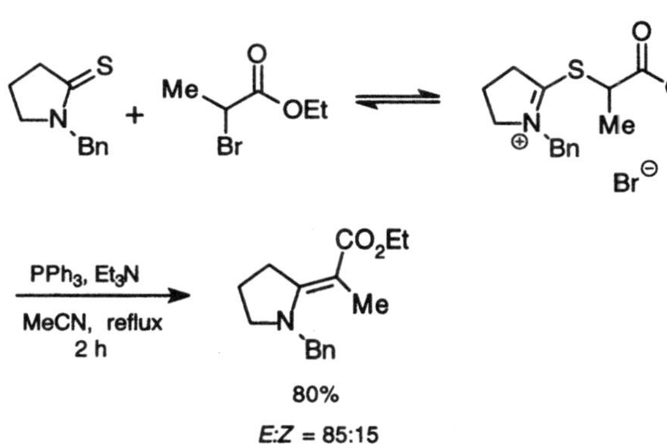

Scheme 48

3.6 Radical Additions

The thiocarbonyl group is excellent for radical addition, which takes place on the sulfur atom and leads to a carbon-centred radical stabilised by the α-sulfur atom. The Barton reaction has enjoyed a great many applications. It mainly involves xanthates and provides many useful processes, such as deoxygenation, decarboxylation, addition to multiple bonds, etc. A number of reviews by Crich et al. have appeared [188, 189], and the most recent is due to Zard [190].

For the compounds that we overview here, an attractive conversion of thionoesters to ethers [191] has used the radical addition of an excess of triphenyltin hydride, instead of tri-*n*-butyltin hydride, which avoided the deoxygenation reaction. The process involves in a first step the very favourable formation of a sulfur-tin bond and the subsequent desulfurisation of the mixed thioketal.

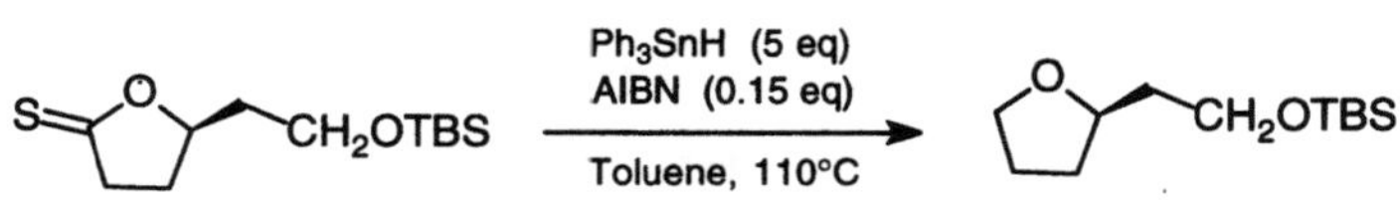

Scheme 49

Thioformamides are reactive towards tin radicals and this was illustrated by a total synthesis of (–)-α-kainic acid via an intramolecular radical cyclisation and further manipulations [192].

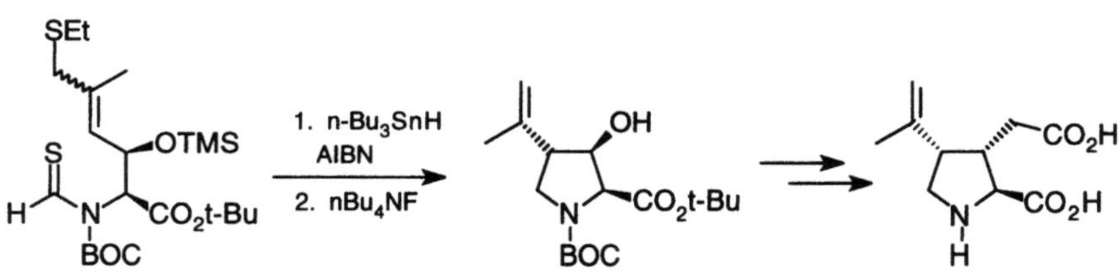

Scheme 50

Another report has dealt with thioamides involved in radical addition [193]. Substrates bearing an α-cyclopropyl group were transformed into cyclopentyl thioamides by thiophilic addition of a trimethytin radical, opening of the three membered ring, addition of the formed radical to an alkyl acrylate, and cyclisation.

Structural investigation of the intramolecular addition of aryl radicals to a dithioester group, in analogy to the carbanionic thiophilic addition reported above, has shown that 6-, 8- and 15-membered rings are accessible in this manner [142].

3.7 [3,3] Sigmatropic Rearrangements

The sulfur analogue of the Claisen rearrangement is the subject of much attention. Molecular orbital calculations have been carried out [194] to compare the oxygen, nitrogen, phosphorus and sulfur versions of this transposition and their transition states. From bond lengths, angles and energies it follows that the sulfur [3,3] shift proceeds more readily than the oxygen one. Activation energies are smaller, but reaction enthalpies are larger. Thus the sulfur rearrangement is *kinetically favoured* [149] and thermodynamically less favoured.

Ab initio calculations [195] have been carried out to model the effect of some substituents: substitution on the C2 carbon by a π donor group stabilises the product more than the starting material and thus will be favourable for the reaction enthalpy. An NH_2 group will provide more stabilisation than SH. This correlates well with the experimental results.

Scheme 51

The starting materials are usually thiocarbonyl compounds, and among them thioamides are the most popular. Two routes allow access to *S*-allylketene aminothioacetals from thioamides: deprotonation by base followed by *S*-allylation, or vice versa.

Meyers and Devine [196] have used the first one: formation of the enethiolate with LDA and addition of an allylic bromide provided the ketene aminothioacetal. In other cases this compound is usually not observed, but here, due to steric hindrance, it was detected, and the facility of the rearrangement, as compared to the oxygen series, was illustrated by the fact that it required only stirring at *room temperature* to occur. An excellent stereoselectivity of 91:9 was achieved for this creation of two new asymmetric centres, including one or even two quaternary ones. The synthetic value was demonstrated by the transformation into enantiopure cyclohexenones.

A second route involves *preforming* first the allylic halide addition to form the iminium salt arising from the alkylation on the highly nucleophilic sulfur atom. Subsequent treatment with triethylamine provided the *S*-allyl thioimidoester which was usually not observed but underwent a deprotonation and a fast [3,3] sigmatropic transposition to give the α-allyl thioamide. With pyroglutamate derivatives no diastereoselectivity was found. An analogous sequence [197] was used with a thioacyl proline derivative, with some stereoselectivity.

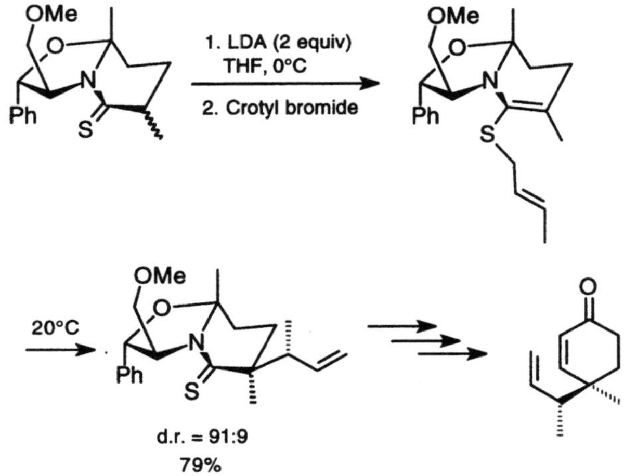

Scheme 52

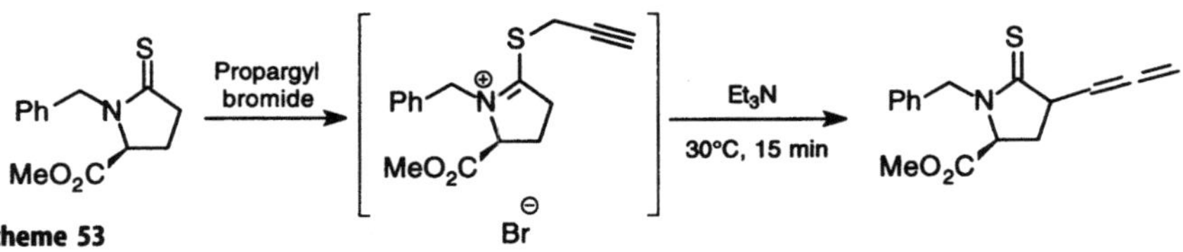

Scheme 53

The thio-Claisen rearrangement can also be performed with secondary thioamides, providing allylation on the nitrogen atom rather than carbon atom. An elaborate example was reported by Fuchs and Smith [183]. They introduced an allyl ammonium salt to achieve SN_2' alkylation of sulfur, and the product readily rearranged to furnish the expected thiolactam. However they finally preferred to employ in the first step an allyl mesylate, which was more easily synthesised.

The second class of starting materials are dithioesters, due to the facility of their access, deprotonation and *S*-alkylation by allyl halides. Over the last few years, studies have concentrated on acyclic stereocontrol, both relative and absolute. In the oxygen series, studies on the Claisen transposition have dealt with substrates bearing stereochemical elements (asymmetric carbon centres, C=C configurations) as part of the pericyclic nucleus. In contrast, very few reports are available in situations where acyclic stereogenic centres are adjacent to the pericyclic array. The Caen group has examined the diastereoselectivity of the rearrangement with carbon or sulfur at the centre of chirality, using either steric or electronic effects.

A dithioester bearing methyl and *t*-butyl groups on the stereogenic centre was deprotonated and allylated to afford a ketenedithioacetal which was rearranged at 100 °C to provide the desired unsaturated dithioester in a 95:5 diastereoisomer ratio [198]. This is simply explained by the steric bulk of the *t*-butyl group and, although the configuration has not been assigned, a

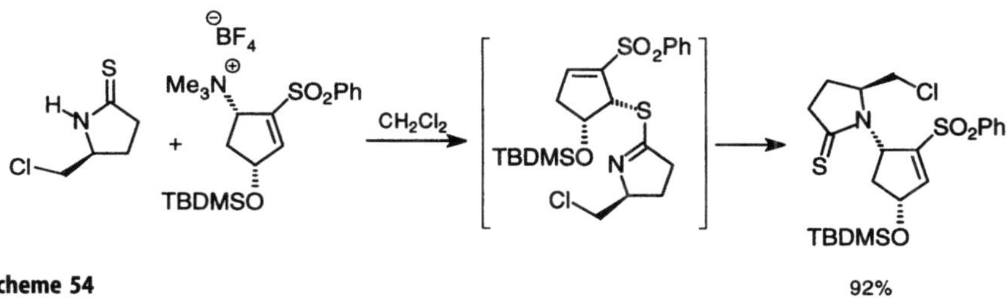

Scheme 54

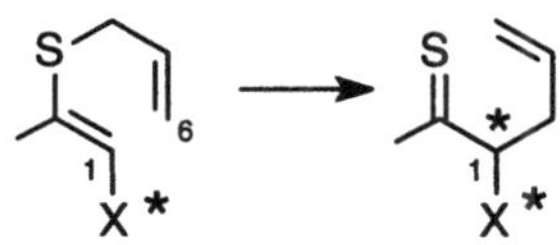

Scheme 55

model is proposed. On the other hand such control was not observed [199] when an asymmetric centre with similar substitution was appended to the terminal carbon of the allyl chain (carbon 6), leading to a diastereoisomer ratio of 75:25.

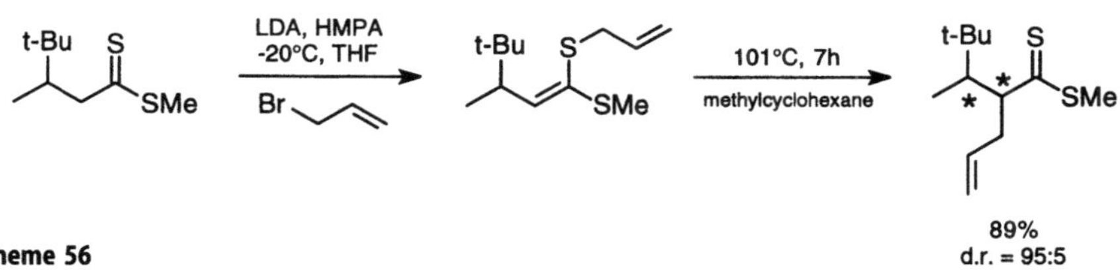

Scheme 56

Beslin and Perrio have pursued their investigation on substrates bearing an hydroxyl and an alkyl group attached to the stereogenic centre, for which they had disclosed [200, 201] a high diastereoselectivity, mainly based on electronic factors. They have succeeded in controlling the formation of three contiguous asymmetric centres. Using (*E*) or (*Z*)-crotyl bromides they have been able to determine the fate of the four possible isomeric ketenedithioacetals in a very sound study.

The noteworthy *cis* selectivity of the deprotonation of thiocarbonyl compounds was used to prepare (*Z*) ketenedithioacetals by starting from *alkyl* dithioesters and *allyl*ating the enethiolates. Conversely, (*E*) ketenedithioace-

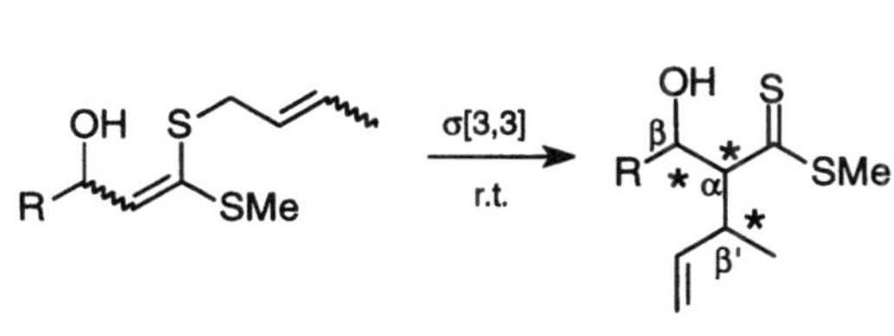

Scheme 57

tals were obtained from *allyl* dithioesters and *alkyl*ation. The rearrangement of the four isomers at ambient temperature leads to four isomeric products in excellent chemical yields (Table 1). In most cases, one out of the four isomers is predominant and the assignment of their relative configurations led to the conclusion that the rearrangement is stereospecific at both double bond configurations of the starting material. Models are proposed to account for the observed selectivities.

Table 1. Thio-Claisen rearrangement

Configurations	
Ketenedithioacetal[a]	Product
ZE'	*anti anti*
ZZ'	*syn syn*
EE'	*anti syn*
EZ'	*syn anti*

[a] The first letter refers to the geometry of the ketene dithioacetal C=C bond and the second to the one of the crotyl moiety.

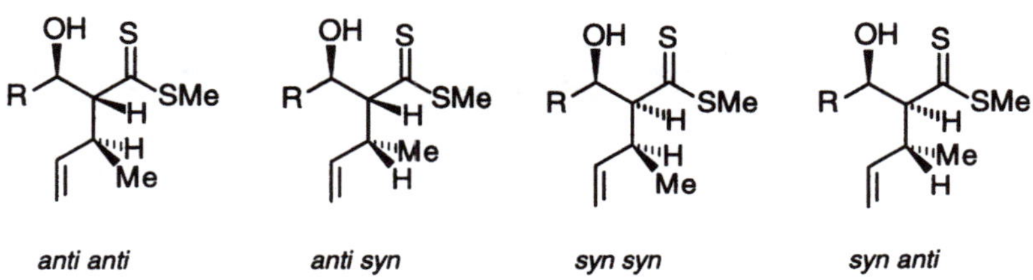

Scheme 59

A different framework was obtained from aldols bearing two asymmetric centres with a *syn* relationship [125]. A similar sequence as above furnished the (*Z*) ketenedithioacetal which rearranged after 2 days at ambient temperature with a high selectivity in favour of the *syn syn* diastereomer. For reasons reported earlier [200, 201], the product configuration is independent of that of the starting material: the (*E*) ketenedithioacetal also leads to the *syn syn* isomer.

A new type of asymmetric induction in this area has been achieved in our group. The need for absolute stereocontrol with the aid of a removable chiral centre, and the observation that the sulfinyl group, recently introduced for Diels-Alder reactions, has not yet been introduced in substrates prone to [3.3] sigmatropic processes, were our motivations. The requisite racemic ketene dithioacetals bearing a sulfinyl group have been prepared and their rearrangement was shown [202] to proceed at ambient temperature. The asymmetric induction was extremely effective, with diastereoselectivities ranging from 93:7 to 99:1, in favour of the (2*S*,*SS*) isomer.

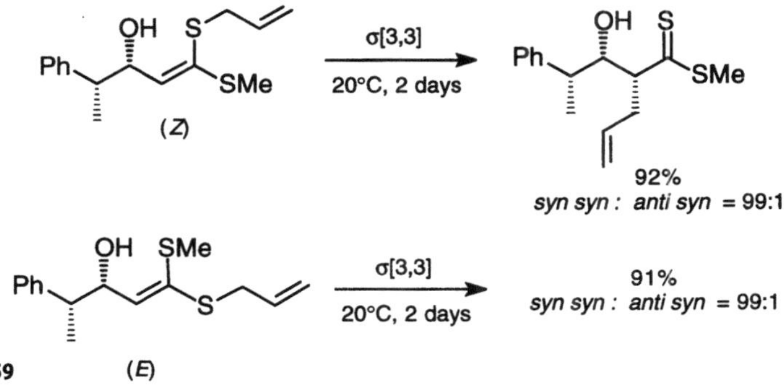

Scheme 59

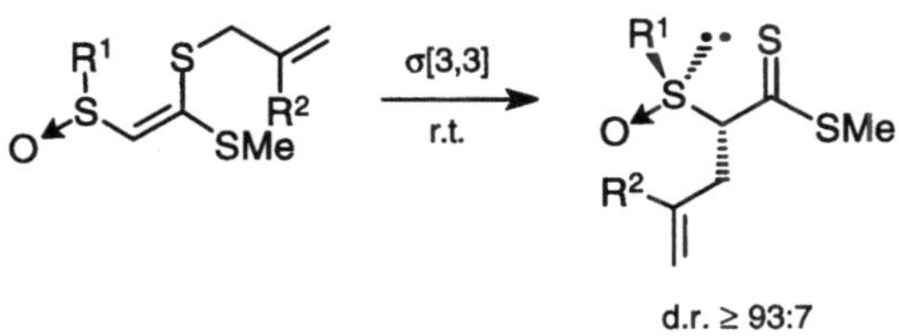

Scheme 60

A model is presented, as a natural extension of the Felkin-Anh one (for nucleophilic addition to an sp^2 centre adjacent to a chiral centre), to explain the preferred formation of one diastereoisomer.

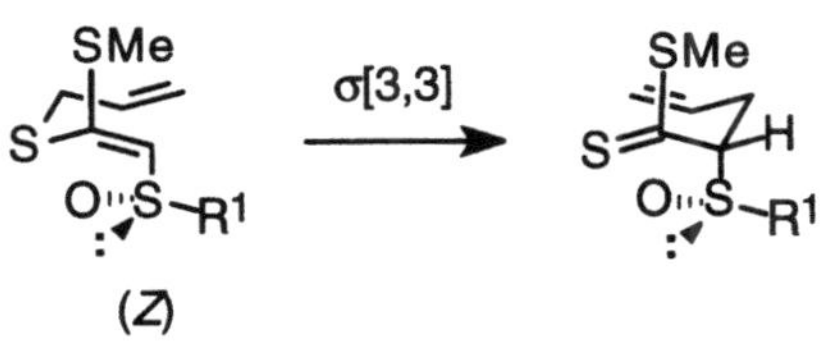

Scheme 61

It is remarkable that elimination of sulfenic acid was not observed during the formation of unsaturated sulfinyl dithioesters. These results are now being extended to enantiopure starting materials, with diacetone glucose as a chiral source of the starting sulfoxides [203].

The generation of unsaturated bis(sulfines) by a dithio-Claisen rearrangement is discussed in Sect. 2.6. The intermediacy of these products is proposed to take place in nature from unsaturated thiosulfinates formed in onion and garlic, and to be the source of zwiebelanes [91–93]. Studies by Block and his group produced a great deal of information about this fascinating chemistry. Activation parameters as low as 15.5 kcal/mol for $\Delta H^{\ddagger}$ have been measured.

Along the same lines, the influence of the sulfinyl group on a sigmatropic shift, Hwu and Anderson [204] have reported on the kinetics of an allyl vinyl sulfide and its analogous sulfoxide. They have observed a 45-fold acceleration due to the oxidation and thus to the electron depletion of the sulfur centre on the 3-position of the pericyclic nucleus. Furthermore, the equilibrium observed with the sulfide (13:87 at 23 °C) was shifted for the sulfoxide in complete favour of the sulfine. Referring to the calculations presented above [194, 195], this might reflect that this rearrangement is *kinetically and thermodynamically* favoured as compared to the oxygen Claisen shift. In many other cases reported by Julia et al. [96–98], and described in Sect. 2.5, the rearrangement is so fast that the unsaturated sulfoxide is not even observed and the sulfines are formed directly.

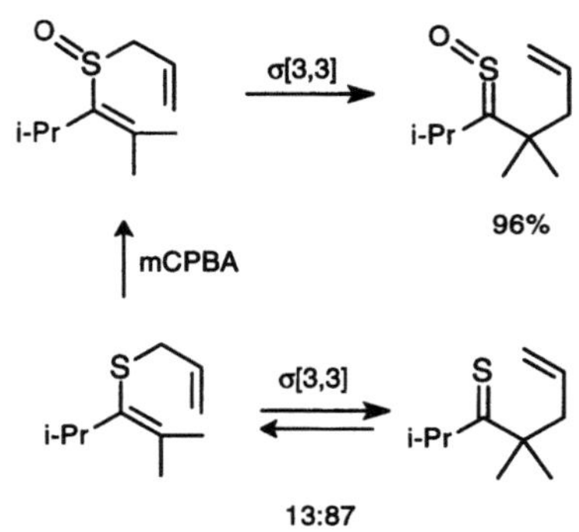

Scheme 62

3.8 [4+2] Cycloaddition Reactions

The last years have seen a great development of the hetero Diels-Alder reactions of thiocarbonyl compounds, due to their rather high reactivity, as reported by many groups, including those of Kirby, Bonini, Capozzi, Degl'Innocenti, Vallée, Okazaki, Koizumi, Saito and Huisgen. A simple orbital frontier treatment allows comparison of the reactivity of C=S vs C=O [119].

A variety of behaviour has been encountered: the thiocarbonyl compounds can act as *dienophiles*, or, when the C=S group is conjugated with a C=C or a C=O double bond, these compounds can react either as *dienes* or *dienophiles*.

3.8.1 *Thiocarbonyl Compounds as Dienophiles*

Kinetic studies [205] have demonstrated that aromatic thioketones are highly dienophilic. Thiofluorenone exhibits an extremely high rate constant (5.12×10^{10} l/mol s) for the reaction with cyclopentadiene and an activation enthalpy as low as 8.2 kcal/mol with 2,3-dimethylbutadiene.

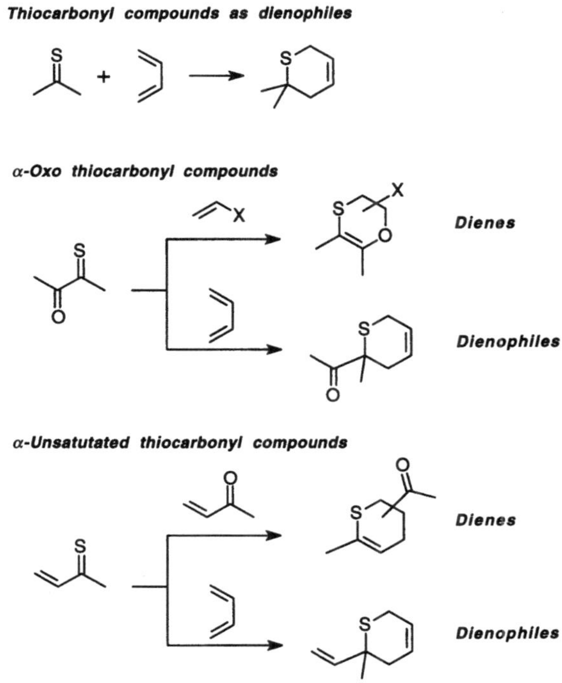

Scheme 63

Stereochemical studies have established [206] that the thermal reaction of thiobenzophenones with (*Z*),(*E*)-2,4-hexadiene leads to dihydro-(2*H*)-thiapyrans with *cis*/*trans* ratios usually superior to 98:2. On the other hand, achieving the addition at 12 kbar and ambient temperature (with a special technique to avoid diene isomerisation) provided mixtures in which the *trans* isomer was predominant. It is remarkable that, with pressure, conservation of the diene geometry with a concerted pathway is observed whereas, at normal pressure, a stepwise mechanism has to be assumed. This unprecedented dichotomy is probably linked to the low reactivity of the *cis,trans* diene isomer.

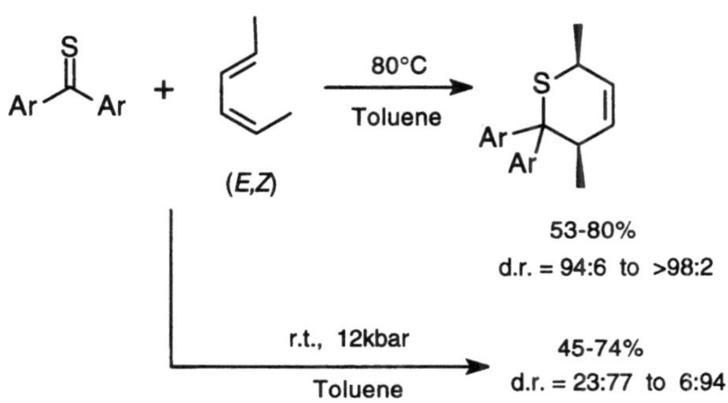

Scheme 64

The series of results are presented as a tabular survey with representative examples (Tables 2 and 3). When the starting thiocarbonyl compounds were not isolated but prepared in situ, their structures appear in brackets.

Diene adducts of *aliphatic* thioketones were prepared, by the group of Degl'Innocenti [75], in a reaction which took place at room temperature (Table 2, entry 1). Even the very elusive cyclohexanethione could be characterised in this manner. A full paper [74] reports many adducts which were prepared by thionation of aldehydes with bis(trimethylsilyl)sulfide and $CoCl_2.H_2O$ in the presence of a 1,3-diene (Table 2, entry 2). Introduction of a trialkylsilicon group on the diene or dienophile was effected by Bonini et al. [80, 207] to facilitate cycloaddition. Reaction of a trialkylsilylenone with thiobenzophenones (Table 2, entry 3), as well as addition of a trialkylsilyl cycloalkyl thioketone with butadiene (Table 2, entry 4), are high yielding processes, taking place at room temperature. Similarly, thioformylsilanes [208] were very recently employed (Table 2, entry 5).

Thiocarbonyl compounds bearing a heteroatom substituent, such as thionoesters, dithioesters or thioamides, are much less reactive. In the glyconolactone series it was shown [67] that a thionester adds to Danishefsky's diene at 110 °C (Table 2, entry 6).

The reluctance of thioamides to be involved as dienophiles led Vallée et al. [209, 210] to use ab initio calculations to model suitable substitution to achieve Diels-Alder reactions. Reduction of the π donor character of the amino group by introduction of a π acceptor group was investigated. *N*-Acylthioformamide indeed reacted at 0–20 °C with 2,3-dimethylbutadiene in the presence of $TiCl_4$ (Table 2, entry 7).

An interesting study was launched by Tamaru et al. [211, 212] with a mono-thiomaleimide for which two cycloaddition reactions with a diene may compete: addition on the C=S or on the C=C double bonds. They found a dependence on the diene structure, such that both processes were encountered (Table 2, entry 8).

The preparation of thioaldehydes and thioketones bearing an adjacent electron-withdrawing group, such as a carbonyl group, has been an intense area of study (Table 3).

In most instances these highly reactive compounds were not isolated, and often not even detected by spectroscopy. They were prepared in situ, by elimination or retro-Diels-Alder reactions, and the trapping agent, such as a diene, was present (and compatible!) at an early stage of the overall sequence. A pioneer of this field, Gordon Kirby, has recently reviewed his contribution [5]. Convenient starting materials are Bunte salts. They were treated by triethylamine to produce, by the elimination of an excellent leaving group, SO_3^{2-}, thioxoacetates as transient species which were trapped by the 1,3-diene, leading to Diels-Alder adducts (Table 3, entry 1).

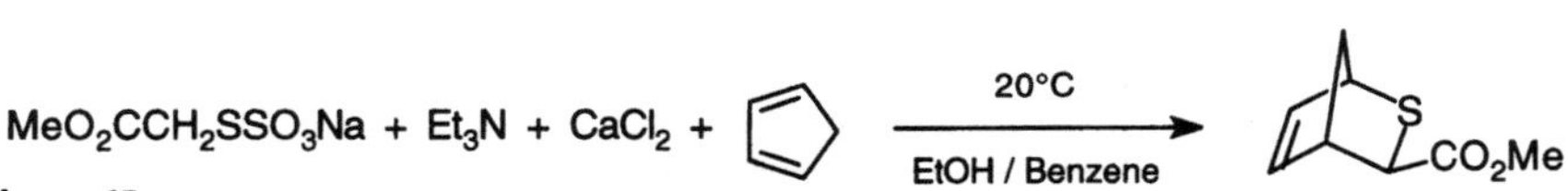

Scheme 65

Table 2. Hetero Diels-Alder reactions of thiocarbonyl compounds as dienophiles

Entry	Thiocarbonyl compound	Diene	Conditions	Product	Yield (%)	Reference
1	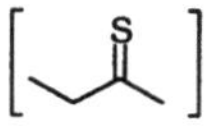	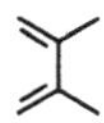	r.t.	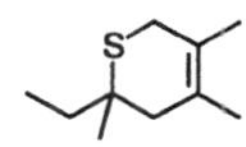	43	[75]
2	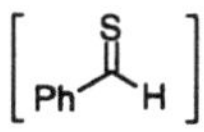	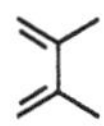		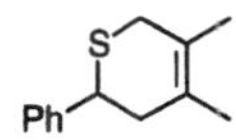	88–90	[74]
3	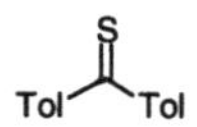	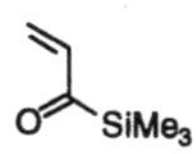	Et_2O r.t.	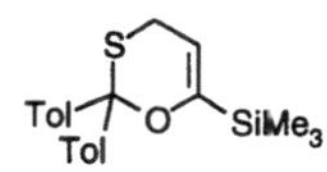	100	[207]
4	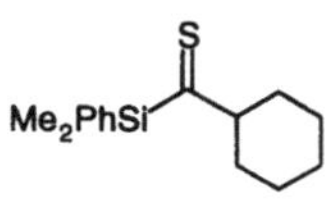		Et_2O r.t.	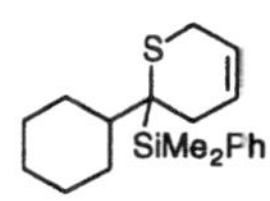	96	[80]
5	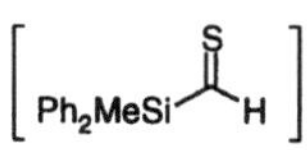	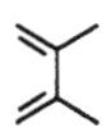	MeCN r.t., 2 h	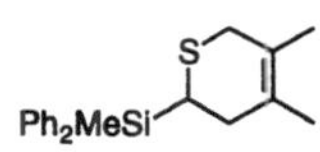	62	[208]
6	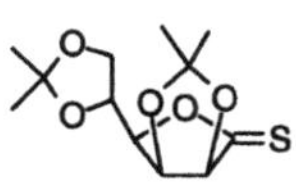	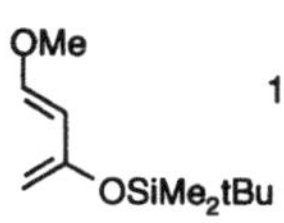	neat 110 °C, 4 h	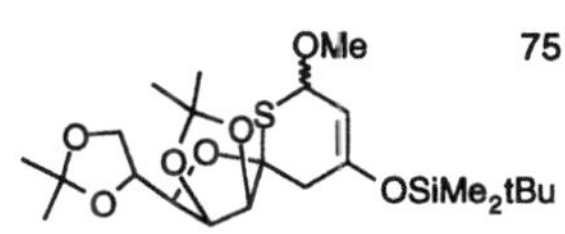 3:1 isomers	75	[67]
7	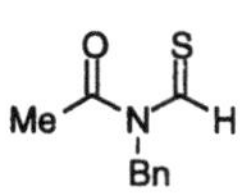	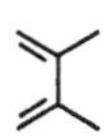	$TiCl_4$ CH_2Cl_2 0-20 °C	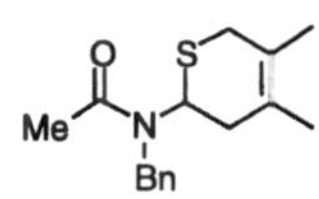	60	[209] [210]
8	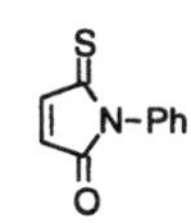		Benzene r.t., 48 h	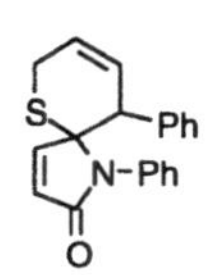	86	[211] [212]
			Benzene r.t., 48 h	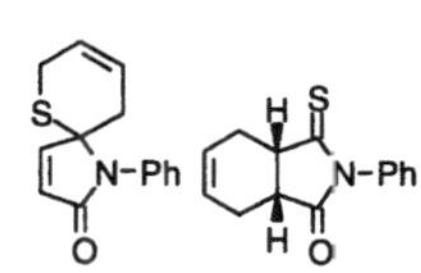 45:55	100	

Table 3. Hetero Diels-Alder reactions of α-oxo-thiocarbonyl compounds as dienophiles

Entry	Thiocarbonyl compound	Diene	Conditions	Product	Yield (%)	Reference
1	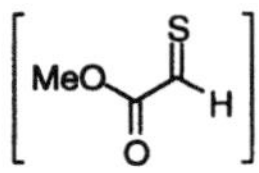	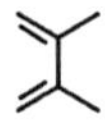	Benzene r.t.	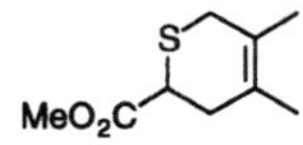	85	[214] [213] [5]
2	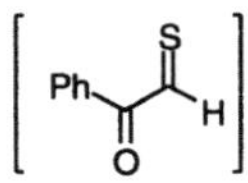	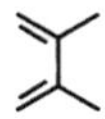		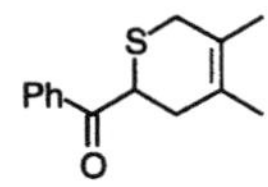	85 80	[215] [216]
3	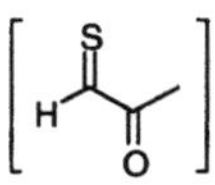		2 equiv $(Me_3Si)_2S$	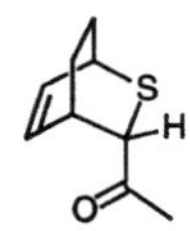 *endo/exo* = 97:3	85	[74]
			1 equiv $(Me_3Si)_2S$	*exo* only		
4	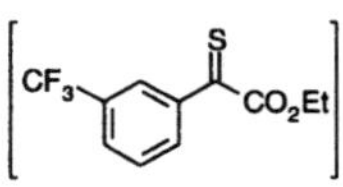		EtOH	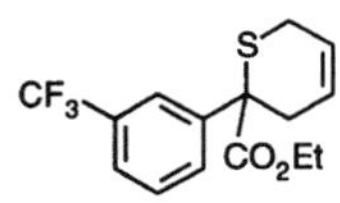	65	[218]
5	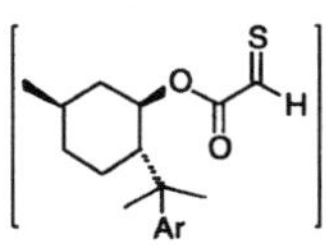 Ar = 4-t-Bu-C_6H_4		THF r.t., 3 h	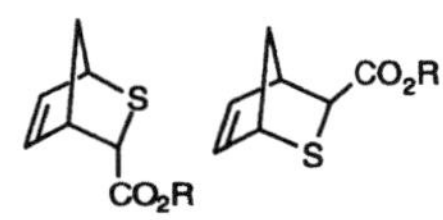 *endo* de = 43% *exo* de = 46% *endo:exo* = 64:36	61	[219]
6	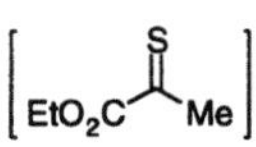	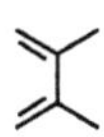		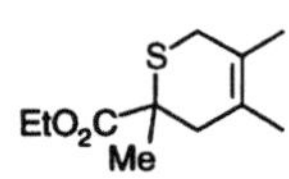	75	[220]
7	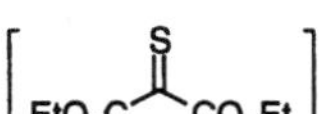	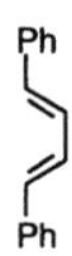	MeCN r.t.	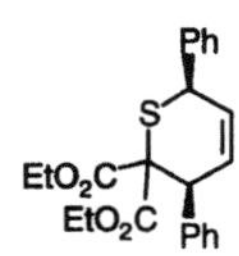	91	[221]

Heating these products (*endo* and *exo* mixtures) was also a source of thiocarbonyl compounds, by cycloreversion [213], and led, in the presence of a diene, to different adducts (*exo* rich equilibrium mixtures). These reactions were applied in the field of morpholine alkaloids and thioshikimic acid synthesis [5]. Another effective source of thioaldehydes are sulfenyl phthalimides [214], which are very prone to elimination, as described hereunder in the work of Capozzi.

Analogous chemistry was reported [215] with thioaldehyde intermediates generated from a non-classical thermal reaction of a sulfoxide via a sulfenate ester (Table 3, entry 2). The reaction of cyclic tetra- or pentasulfides with stabilised phosphoranes is also a good source of thioaldehydes [216], trapped by dienes (Table 3, entry 2).

The thionation of aldehydes with bis(trimethylsilyl)sulfide and $CoCl_2.H_2O$ in the presence of a 1,3-diene was applied to pyruvaldehyde (Table 3, entry 3). In the presence of TMSOTf a remarkable stereochemical effect was observed: *one* equivalent of the Lewis acid leads to a 97:3 *endo:exo* mixture, whereas *two* equivalents furnish selectively the *exo* isomer [74].

The facility of these cycloadditions was exploited in the field of the synthesis of a new class of drugs: potassium channel activators, promising for the treatment of hypertension [217]. After discovery of aprikalim by Rhône-Poulenc (Sect. 2.1), various methods of synthesis as well as preparation of many analogues have been investigated. Chemists at SmithKline-Beecham have reported [218] a novel strategy involving a cycloaddition route (Table 3, entry 4). An α-thioxo-ester was generated from a Bunte salt in basic medium and added to various dienes to afford good yields of dihydrothiapyrans, on which other transformations were carried out (oxidation to the sulfoxide, conversion of the ester into an amide).

Koizumi and his group have published [219] the first attempt to achieve an asymmetric version using 8-arylmenthols as chirality auxiliaries. Enantiopure methyl thioxoesters were prepared by reaction of dichloroacetates with $(Bu_3Sn)_2S$/TBAF. Cycloaddition to cyclopentadiene led to encouraging results (Table 3, entry 5).

Thioxomalonates and α-thioxopropanoates have been prepared and subsequently trapped [220] with various dienes [221] (Table 3, entries 6 and 7). The resulting dihydro-2*H*-thiapyrans have been exposed to a strong base to effect a ring contraction providing a new synthesis of cyclopentenes [220]. Other adducts led to the formation of vinyl cyclopropanes, which are possible intermediates in cyclopentene synthesis.

3.8.2
Thiocarbonyl Compounds as Heterodienes

Conjugation of a C=S moiety with a C=C or a C=O bond brings the possibility of the compounds playing the role of dienes in cycloaddition (Tables 4 and 5). This is indeed observed with aromatic thioketones as demonstrated by the extensive work of Saito et al. at the Science University of Tokyo. Reaction of furyl phenyl thioketones or thiophenyl phenyl thioketones with

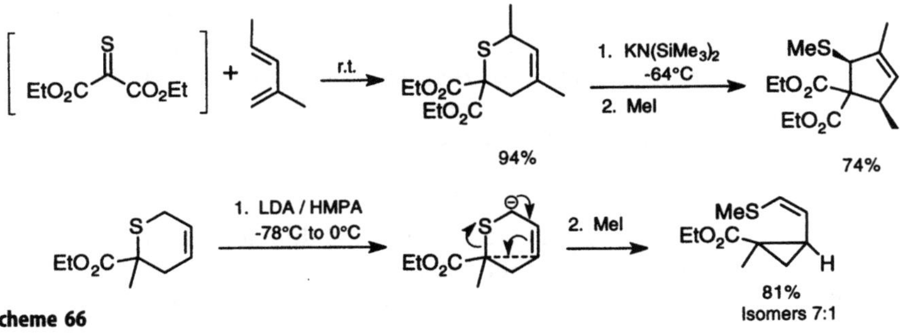

Scheme 66

α-chloroacrylonitrile leads [222] to cycloadducts (Table 4, entry 1) arising from the addition of an alkene to an unsaturated thioketone, with the C-C double bond involved being that of the heteroaromatic ring (furane or thiophene).

In search for control of absolute stereochemistry, the reaction of thiochalcones was investigated with unsaturated amides bearing an Evans chiral oxazolidinone [223] and dimenthyl fumarate [224, 225]. For the first time with thiocarbonyl compounds, the efficiency of Lewis acid addition was demonstrated, and reactions could be conducted at room temperature. With $EtAlCl_2$ (Table 4, entry 2) or $AlCl_3$ (entry 3), levels of induction up to 92% were attained for the *endo* isomer. $Yb(OTf)_3$ in DMSO also caused the acceleration of the reaction with chiral acrylamides with p-facial selectivity [226]. This group has also reported [227] an intramolecular hetero Diels-Alder reaction with divinyl thioketones and the double bond of an allyloxy group (Table 4, entry 4).

The reaction of thiobenzophenones with electron poor acetylenes [228] takes place at ambient temperature and provides excellent yields of 2*H*-thiapyrans (Table 4, entry 5). Benzyne, whose reaction with C=S compounds was reviewed recently [229], was treated with various thiobenzophenones [230]: a [4+2] cycloaddition was followed by a hydrogen shift (entry 6).

Conjugated thioamides are normally not good partners. Fishwick et al. have nicely exploited the idea that the introduction of an electron withdrawing acyl group on the nitrogen of thioamides should facilitate the reaction. They have recently published [231] a route for the synthesis of enantiopure thiapyrans starting from an unsaturated thioamide bearing on nitrogen both an acetyl group and (*R*)-1(1-naphthyl)ethylamine as a chiral auxiliary. With cyclopentene an *exo* selective addition and a d.e. >98% was achieved (Table 4, entry 7).

In the same field Guingant and Pradère have developed an asymmetric cycloaddition of 3-aza-1-thiabutadienes. The chiral auxiliary was brought here by the dienophile which is an unsaturated Evans amide [232]. It is remarkable that *both* enantiomers were selectively obtained by proper choice of the conditions: catalysis of $MgBr_2$ at 0 °C or a pressure of 10 kbar at 20 °C (Table 4, entry 8).

Table 4. Hetero Diels-Alder reactions of unsaturated thiocarbonyl compounds as dienes

Entry	Thiocarbonyl compound	Dienophile	Conditions	Product	Yield (%)	Reference
1	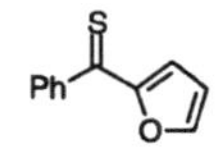	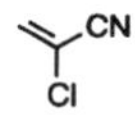	1. benzene reflux 10 min 2. Et_3N, r.t.	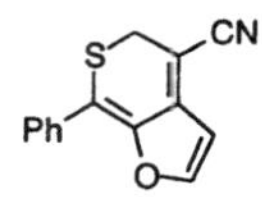	63	[222]
2	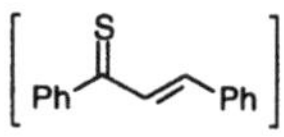	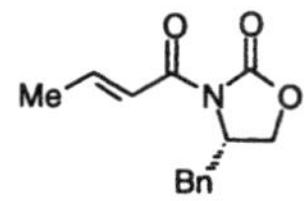	$EtAlCl_2$ (1.0 equiv) CH_2Cl_2 25 °C, 0.7 h	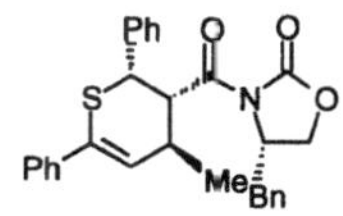 d.e. (*endo*) = 82% *endo* : *exo* = 87 : 13	96	[223] [226]
3	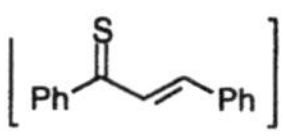	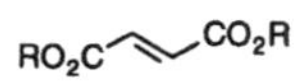 R = (-)-Menthyle	$AlCl_3$ (3 equiv) Et_2O 35 °C, 5 h	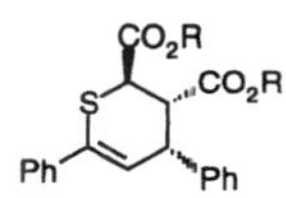	76	[224] [225]
4	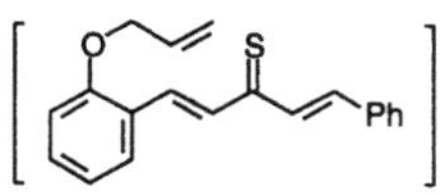		Benzene 80 °C, 0-3 h	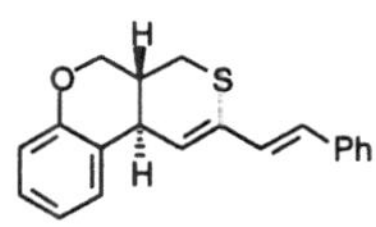 *trans* : *cis* = 90 : 10	85	[227]
5	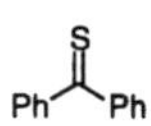	$MeO_2C—≡—CO_2Me$	$CHCl_3$ r.t., 14 d	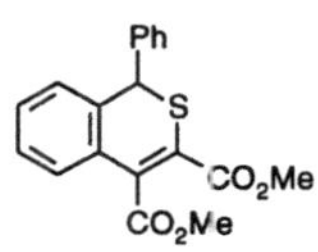	95	[228]
6	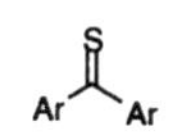 Ar = 4-MeO-Ph	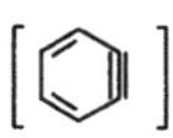	CH_2Cl_2 reflux 3 h	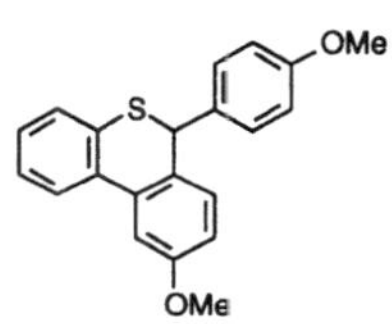	44	[230]
7	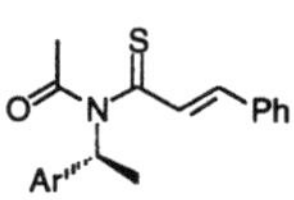 Ar = 1-Naphtyl		r.t.	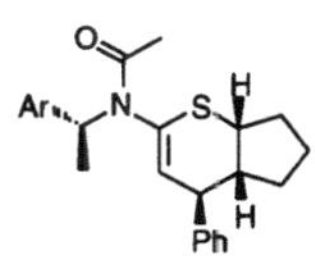 *exo* de >98%	95	[231]

Table 4 *(continued)*

Entry	Thiocarbonyl compound	Dienophile	Conditions	Product	Yield (%)	Reference
8	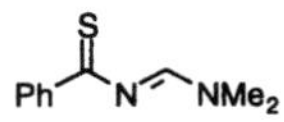	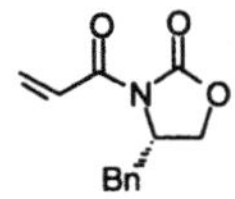	$MgBr_2$, CH_2Cl_2, 0 °C, 3 h	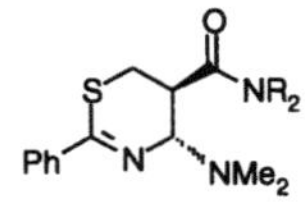 dr = 100:0	95	[232]
			10 kbar, 20 °C, 40 h	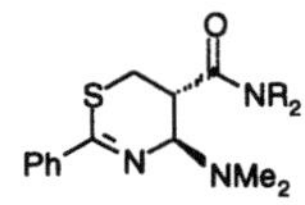 dr = 84:16	95	[232]
9	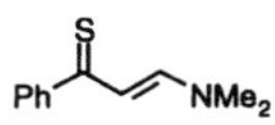	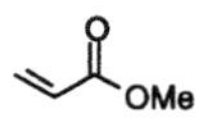	CH_2Cl_2, -30 °C, 1.5 h	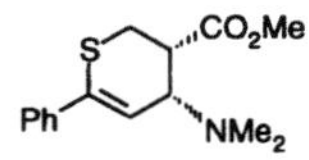 *endo* : *exo* = 91:9		[233]
			80 °C, 3 h	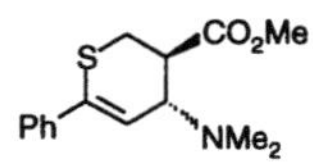 *endo* : *exo* = 0:100		
10	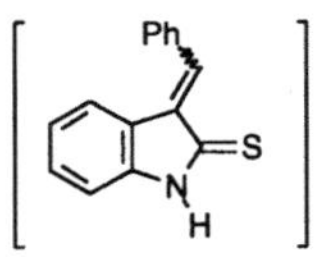		20 °C	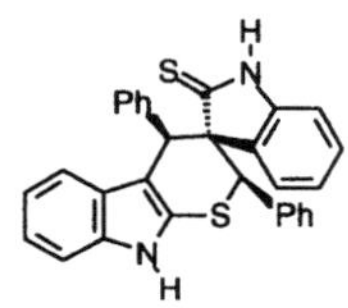	82	[234]
11	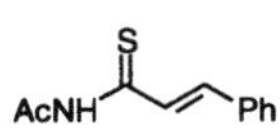	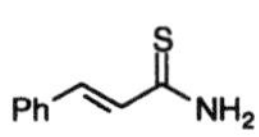	Pyridine, Acetone, reflux, 3 h	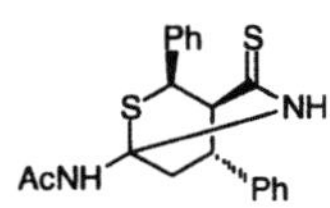	39	[235]
12	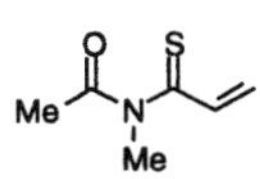	[60] Fullerene	Pyridine, 65 °C, 30 min	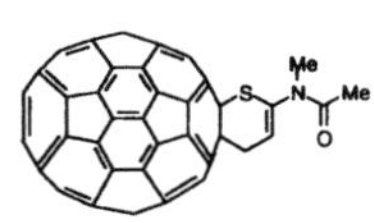	57	[236]

Other results with unsaturated thioamides include a nice *reversal* [233] of the *endo* selectivity in favour of the *exo* adduct by adjustment (Table 4, entry 9) of the thermal conditions (–30 °C vs 80 °C). Unexpected dimerisations of enethioamides (entries 10 and 11) were observed [234, 235]. Thiocarbonyl chemistry has already been used in fullerene chemistry Acyl thioacrylamides react [236] with [60]fullerene to yield hetero-Diels-Adler products (Table 4, entry 12).

The third class of conjugated thiocarbonyl derivative used in these reactions are α-oxothiones (Table 5). A novel and extensive investigation was

Table 5. Hetero Diels-Alder reactions of unsaturated thiocarbonyl compounds as dienes

Entry	Thiocarbonyl compound	Dienophile	Conditions	Product	Yield (%)	Reference
1	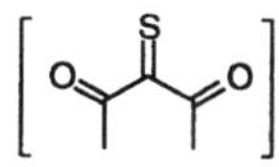	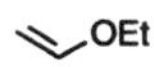	$CHCl_3$ r.t., 2.5 h	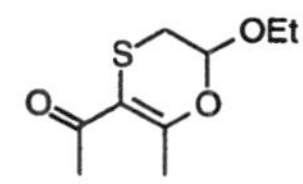	78	[237] [87]
2	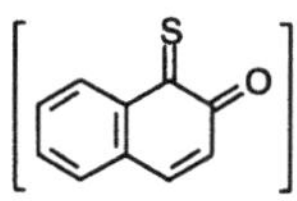	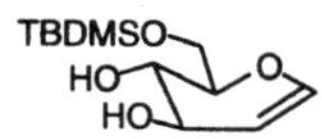	$CHCl_3$ 60 °C	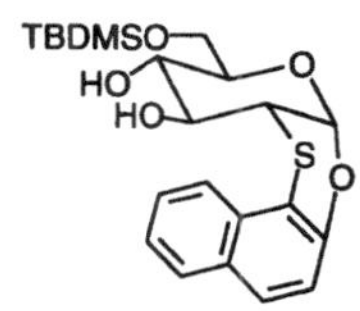	76	[238]
3	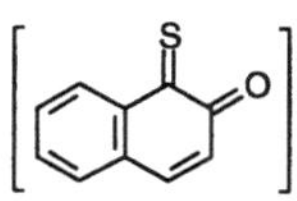	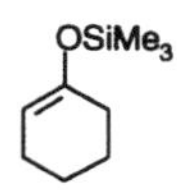	108 h	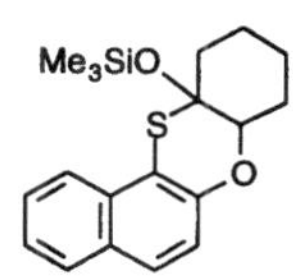	89	[239]
4	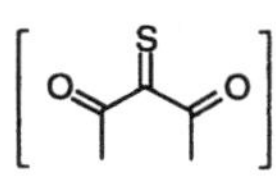	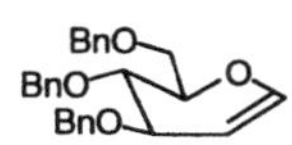	CH_2Cl_2 r.t. $t_{1/2}$ = 3 d	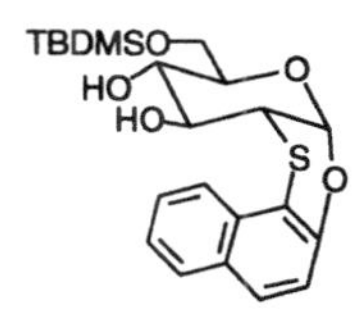 $\alpha:\beta$ = 95:5	80	[240]

launched by Capozzi et al. who demonstrated that their behaviour is not limited to the role of dienophiles, as described above. Their synthesis of starting materials rests on the easy cleavage of sulfenyl phtalimides (SPhth) in basic medium. Thus α,α'-dioxothioketones were transiently formed [87, 237]. They were trapped by electron rich olefins regioselectively to produce dihydrooxathiins (Table 5, entry 1).

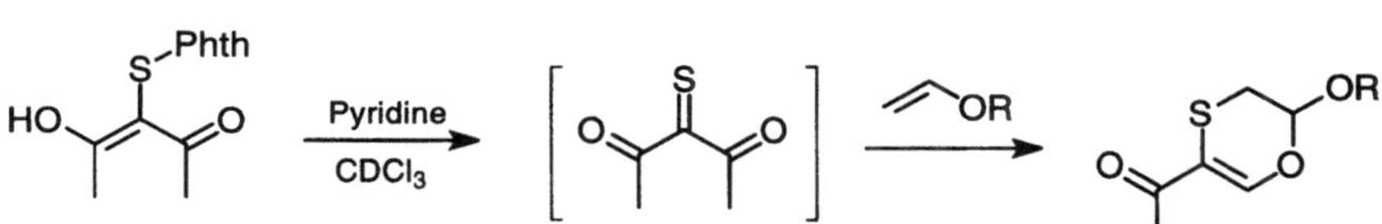

Scheme 67

Ortho-thioquinones are also accessible by this route [238, 239]. They were treated with glycals as dienophiles with success (Table 5, entry 2). These reactions were designed [240] to offer an original glycosyl transfer, by achieving a Diels-Alder reaction with the appropriate α,α'-dioxothione (Table 5, entry 4), and a simple Raney nickel reductive desulfurisation of the cycloadduct.

3.8.3
Sulfines and Cycloadditions

Thioaldehyde *S*-oxides behave as dienophiles with 1,3-dienes [241]. Cycloaddition involves the carbon-sulfur double bond of the heterocumulene to furnish dihydro-2*H*-thiapyran *S*-oxides (Table 6, entry 1). The *cis:trans* product ratio was found to depend upon the initial diene/sulfine ratio. A large excess of diene selectively provided the *cis* cycloadducts, as a result of a stereospecific reaction of the (*Z*) sulfine. With a produced proportion of diene, mixtures of *cis:trans* isomers are obtained, with a preference for the *cis* compound. The Italian authors have demonstrated that the slower reaction allows a (*Z*) to (*E*) isomerisation of the sulfines prior to addition.

A cyanosulfine [242] and an oxosulfine [162] were trapped with 2,3-dimethyl-1,3-butadiene (Table 6, entries 2 and 3). Capozzi and his group have extended their phthalimido-sulfenyl chemistry to the synthesis of α-oxosulfines, and have observed a dichotomic behaviour towards cycloaddition. With 1,3-dienes, these sulfines act [243, 244] as dienophiles through their C=S bond (Table 6, entry 4) to afford dihydro-2*H*-thiapyran *S*-oxides. With alkenes (Table 7), such as 2,3-dimethyl-2-butene (entry 1) or vinylethers (entry 2), they behave as dienes to give dihydro-1,4-oxathiin *S*-oxides [243–245].

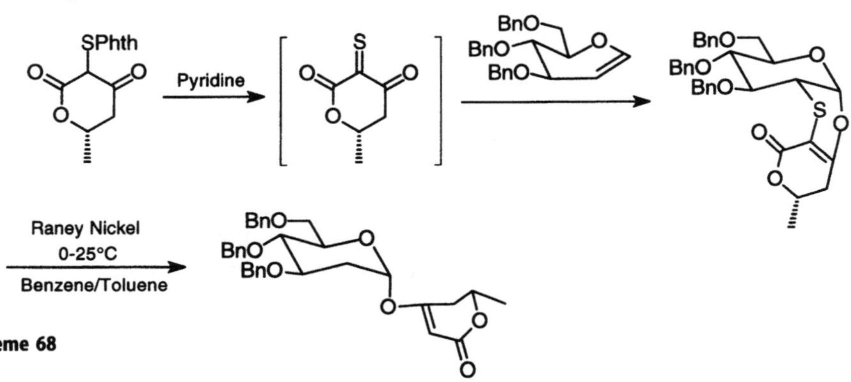

Scheme 68

3.9
Ene Reaction

The ene reaction has not been much studied. Three reports of *intramolecular* versions have however recently appeared (Table 8). The original feature of the thiocarbonyl version is that two pathways are possible: addition to *sulfur* or to *carbon* atoms. Both have indeed been encountered. Kirby and his group [246, 247] have generated allyl thioxoacetates which, depending on the substitution of the C=C double bond, led selectively to a mercapto-lactone (Table 8, entry 1) by formation of a C-C bond, or to an 8-membered ring (Table 8, entry 2) by generation of a C-S bond. The latter process was also observed [248] by the group of Saito with allyloxythiobenzophenones (Table 8, entry 3).

Table 6. Hetero Diels-Alder reactions of sulfines as dienophiles

Entry	Thiocarbonyl compound	Diene	Conditions	Product	Yield (%)	Reference
1	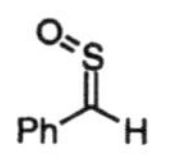 Z:E = 99:1	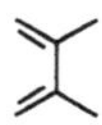	25 °C, 70 h	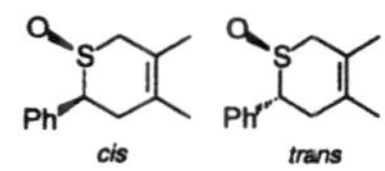		[241]
			7.1 equiv of diene	33:67	60	
			111 equiv	95:5	90	
2	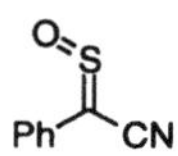	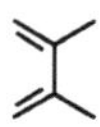	CCl_4 r.t., 72 h	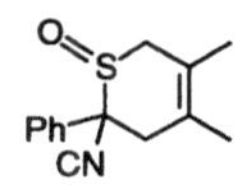	84	[242]
3	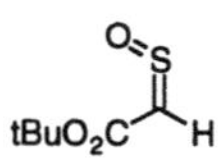	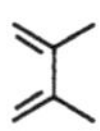		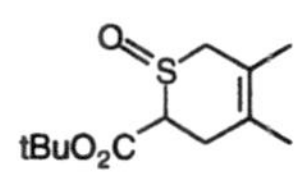	40	[162]
4	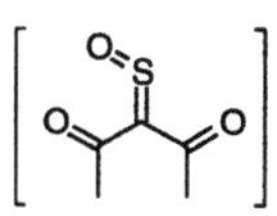	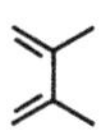	$CHCl_3$ 60 °C	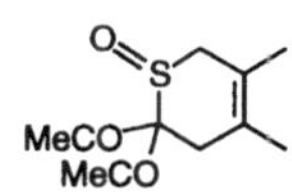	98	[243]

Table 7. Hetero Diels-Alder reactions of sulfines as dienes

Entry	Thiocarbonyl compound	Diene	Conditions	Product	Yield (%)	Reference
1	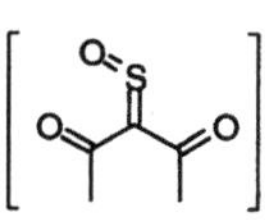	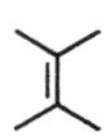	60 °C, $CHCl_3$	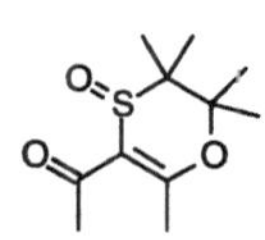	57	[243] [244]
2	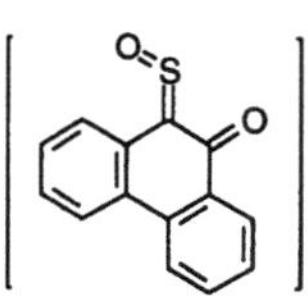	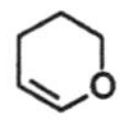	CCl_4 r.t., 24 h	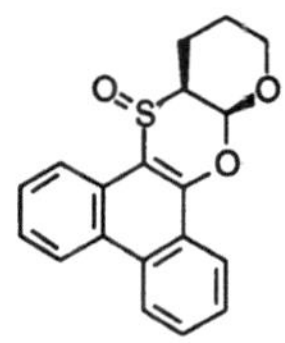	70	[245]

Table 8. Intramolecularene reactions of thiocarbonyl compounds

Entry	Thiocarbonyl compound	Conditions	Product	Yield (%)	Reference
1	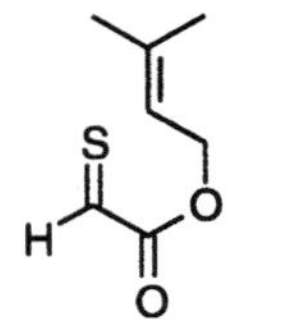 C-C bond formation		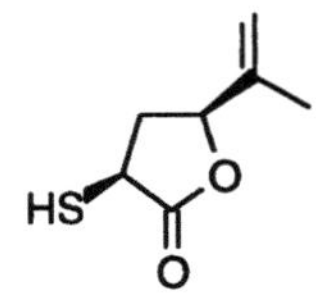	69	[246]
2	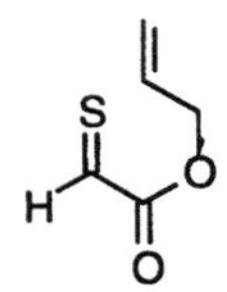 C-S bond formation		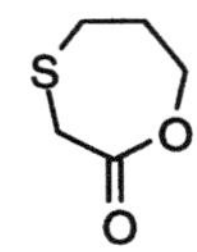	60	[247]
3	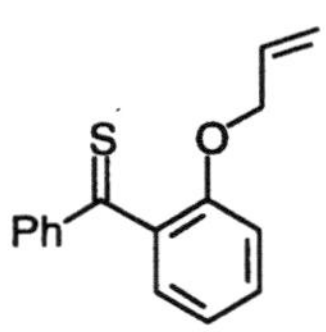	Xylene reflux 0.5 h	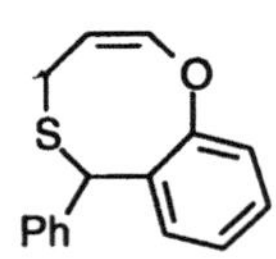	99	[248]

3.10 [2+2] Cycloadditions

Although some examples have previously shown the feasibility of this reaction, very few studies have been reported recently. The isolation of a new antibiotic, leinamycin, featuring a novel dithioperoxyester *S*-oxide moiety, led Pattenden and Shuker [249] to a strategy based upon the [2+2] cycloaddition of a thioketone and a ketene. The expected 4-membered ring thiolactone was obtained, albeit in modest yield, very probably due to the low thermal stability of thioacetone. The product could be transformed into a model dithiolane oxide.

[2+2] Cycloaddition reactions of thioaldehydes and sulfines are most probably encountered in plants, as elegantly and soundly shown by the group of Eric Block during their investigation of sulfur products occurring in the *Allium* species (for a review see [91]). They were able [92, 93] to isolate bicyclic dithioacetal oxides, called zwiebelanes, and also to synthesise them from a thioxosulfine, already described in this review (Sect. 2.6, Scheme 18). An extremely rich stereochemical and analytical study has resulted.

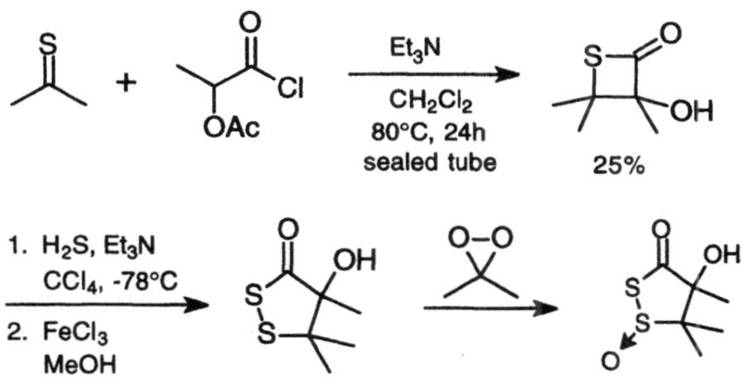

Scheme 69

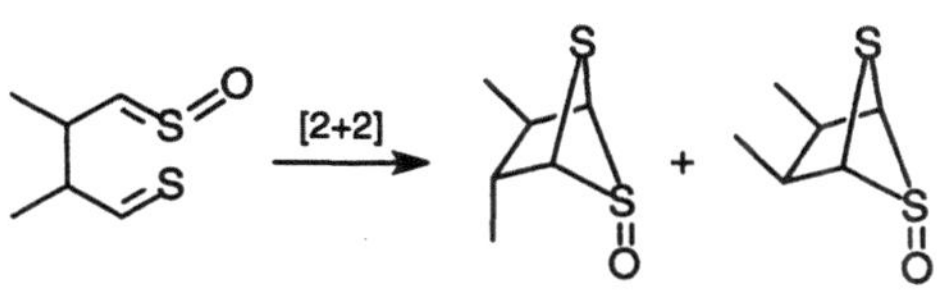

Scheme 70

3.11 Dipolar [3+2] Cycloadditions

A very recommended reading is a review [250] by the master of this field, Rolf Huisgen, and co-workers. They highlight specific features of thiocarbonyl compounds:

- Thioketones are active dipolarophiles; kinetic studies reveal the great superiority of C=S bonds over C=C
- Thione *S*-oxides and *S*-sulfides are dipolarophiles, but they were recently shown also to act as dipoles.

3.11.1 *Thiocarbonyl Compounds as Dipolarophiles*

3.11.1.1 Reaction with Azides

This reaction was intensely investigated by Mloston et al. In a classical, but not general reaction, they observed that heating aromatic or aliphatic thioketones [251, 252], thionoesters [253] and dithioesters [252, 253] with aryl- or benzylazides led to the formation of a carbon-nitrogen double bond. A [3+2] cycloaddition was assumed, with successive elimination of nitrogen and sulfur.

Scheme 71

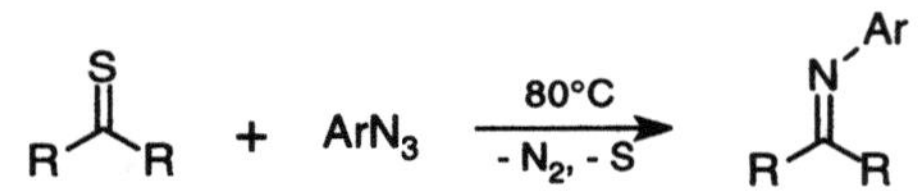

An original observation was made with the sterically hindered 2,2,4,4-tetramethyl-3-thioxocyclobutanone and arylazides [254]. Dispiro-1,3,4-dithiazoles were obtained in 67–83% yields. Their formation was explained by a 1,3-dipolar cycloaddition, elimination of N_2 and formation of a novel thiocarbonyl ylid, which underwent dipolar cycloaddition with cyclobutanethione. The other outcome possible for the ylid was cyclisation to a thiaziridine and sulfur extrusion to produce the imine, as previously reported.

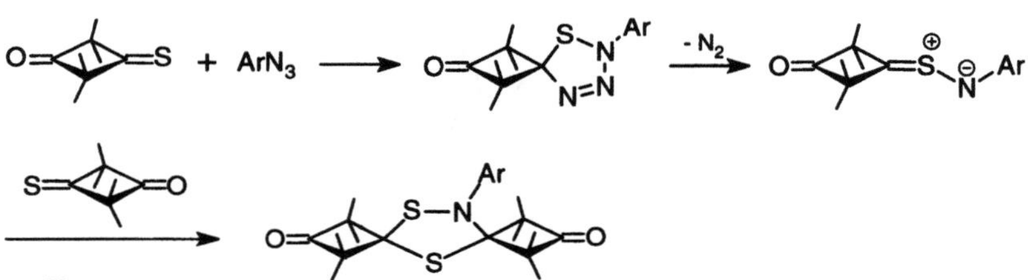

Scheme 72

Thiocarbonyl compounds have been utilised in the attractive concept of tandem *three component reactions.* Proper design could, in principle, allow the use of two different thiocarbonyl compounds in a sequence where the first would generate the ylid and the second would trap it. A success came [253] for a mixture made of the preceding cyclobutanethione, methyl dithiobenzoate and phenylazide, leading in the expected tandem reaction to dithiazole, but in a modestly selective fashion. An interesting trithiolane was also formed, the source of which might be a thione *S*-sulfide. Another *three component reaction* was conceived to generate trithiolanes in good yields. Heating a mixture of cyclobutanethione, 4-dimethoxythiobenzophenone (2 equivalents) and phenylazide furnished the expected trithiolane in a remarkable 69% yield. Among the possible pathways, the author proposes a cycloaddition of the aromatic thione, elimination of N_2, formation of a thiaziridine, reaction of the second equivalent of aromatic thione to furnish both the thione *S*-sulfide and the imine, and lastly dipolar addition of the thione *S*-sulfide to cyclobutanethione. An analogous system was reported with one of the thiones being replaced by fumarodinitrile [255].

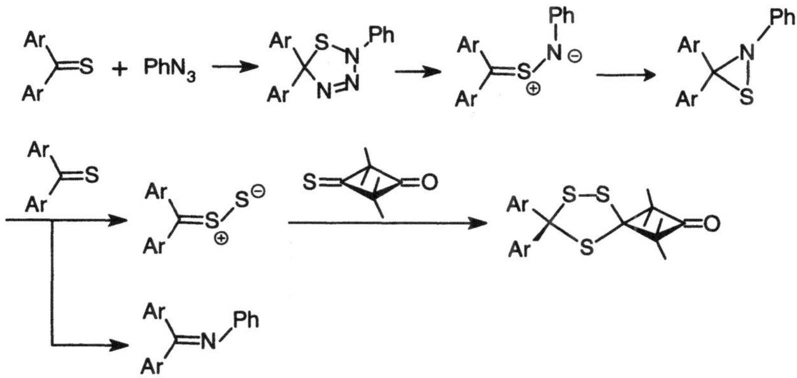

Scheme 73

3.11.1.2
Reaction with Nitrones

A full report by Huisgen et al. [256] describes a number of rate measurements. Reaction of the hindered 2,2,6,6-tetramethylcyclohexanethione with a nitrone has extreme values: low activation enthalpy ($\Delta H^{\ddagger}$ = – 10.8 kcal/mol) and high entropy ($\Delta S^{\ddagger}$ = –28 e.u.).

An outstanding comparison was made with the best known dipolarophile, dimethyl acetylenedicarboxylate (DMAD). The cycloaddition of *N*-methyl-*C*-phenylnitrone to adamantanethione is 1500 times faster than the addition to DMAD. As indicated in Scheme 74 the reaction is an equilibrium, leading to 56:44 ratio of product vs starting material. This equilibrium is shifted towards the left side for thiobenzophenones, as the cycloadduct formation breaks the conjugation, and thus these thiones do not "appear" to react with nitrones. Ab initio calculations were carried out to model the high reactivity of nitrones with thiocarbonyl compounds [257].

3.11.1.3
Reaction with Diazoalkanes

After accumulating a wealth of information on this cycloaddition, the pace of investigation has slowed down. This reaction has seen synthetic application in the sugar series [67]. A thionolactone was reacted with CH_2N_2 to produce a regioisomeric mixture of dihydrothiadiazoles, which can undergo transformations to give thiadiazoles by sugar ring cleavage.

Heimgartner and his group examined the behaviour of thioketones and 1,3-thiazole-5-(4*H*)-thiones towards α-diazoketones at 50–90 °C. A selective formation of 1,3-oxathiole was achieved, as a result of a [3+2] cycloaddition, subsequent nitrogen elimination to produce an acyl thiocarbonyl ylide, and 1,5-dipolar electrocyclisation.

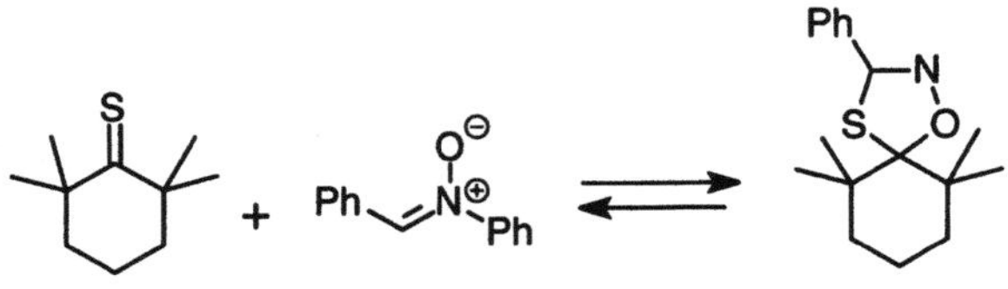

Scheme 74

3.11.1.4
Reaction with Carbonyl Oxides

The isomers of the now very popular dioxiranes, carbonyl oxides, are much less well-known species. However, the reaction of both isomers with two aliphatic thioketones was investigated [258]. A clear cut difference in reactivity was observed, with dioxiranes leading expectedly to the sulfines, while 5-membered ring thio-ozonides were produced from the reactions with carbonyl oxides, thus proving the possibility of a dipolar cycloaddition.

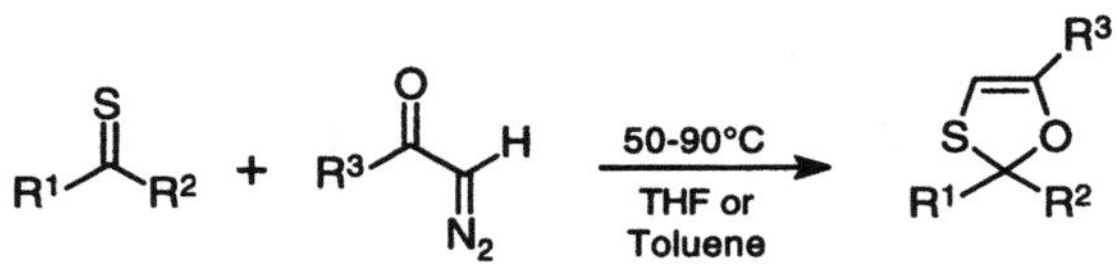

Scheme 75

3.11.2
Sulfines as Dipolarophiles

A precursor of a thiocarbonyl ylid, a dihydrothiadiazole, was reacted [259] with a sulfine, thiobenzophenone *S*-oxide, to lead mainly to a dithiolane *S*-oxide, the formation of which was rationalised through a 1,3-dipolar cycloaddition of the ylid with the C=S bond of the sulfine. It is interesting to note that an opposite regioselectivity was observed for thiofluorenone *S*-oxide.

3.11.3
Sulfines as Dipoles

Huisgen et al. have reported [260] evidence for sulfines playing the role of dipole in [3+2] cycloaddition. The reaction of thiobenzophenones with the *S*-oxide of a cyclobutanethione proceeds at room temperature to yield crystals of 1,2,4-oxadithiolanes, in a process in which reversibility was demonstrated.

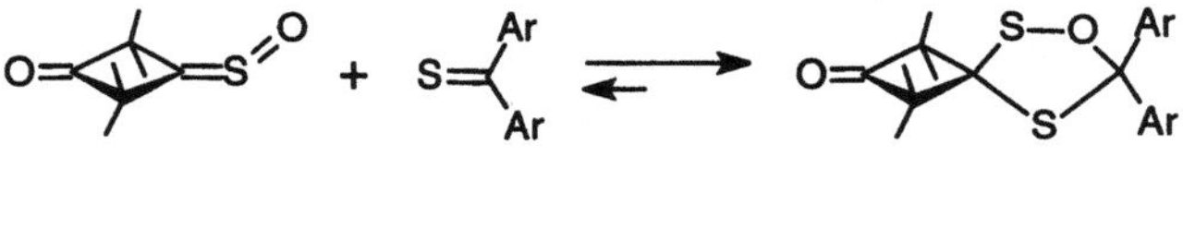

Scheme 76

There is some debate as to whether this is the first report of sulfines as dipoles, and indeed this article mentions two precedents from Block and Schaumann-Walter. The article reports some factors which may help to explain why it took so long to find unequivocal examples of the dipolar activity of sulfines. At the end of this review it is noteworthy and ironic that the solution to this long-standing search was a reaction with a thiocarbonyl compound!

Further examples have since also been found. The reported thermodynamic limitation was circumvented by replacing the aromatic thioketone by an aliphatic one, adamantanethione [261]. The reaction of the latter with 2,2,4,4-tetramethyl-3-thioxocyclobutanone *S*-oxide was achieved at 80 °C to afford the expected oxadithiolane (isolated yield 7%), arising from 1,3-cycloaddition of the sulfine. Products were characterised by an X-ray crystallographic study, and the *"superdipolarophilic"* character of the C=S bond was modelled by quantum calculations. The authors attempted to produce the regioisomer by reaction between a pair of substrates with exchanged functional groups: thiobenzophenone *S*-oxide and the substituted thioxocyclobutanone. Surprisingly, a trithiolane was obtained, instead of an oxadithiolane [262]. A mechanism was proposed, involving cycloreversion of the expected cycloadduct and participation of a thione *S*-sulfide and its dithiirane isomer. In a variant, an unknown dipole, a carbonyl sulfide ($R_2C{=}O^+{-}S^-$), was envisaged, thus proposing new exciting targets for research in the field of dipoles and sulfur chemistry.

4 Conclusion

The last 5 years have seen a great many of new reactions and synthetic applications of thiocarbonyl compounds. This area is far from a routine field of study. Risks have been taken and some have proved fruitful, thanks in general to the high reactivity and selectivity of these compounds and reactions investigated. It is no more the reserved domain of a few specialised groups, as can be seen by the list of references, including a variety of research groups, among them synthetic chemists involved in target molecules with specific biological properties. It is the goal of this overview to encourage the spread of awareness of the results described and to inspire the initiation of new investigations.

Acknowledgements. I wish to thank warmly Dr Carole Alayrac, Dr Stéphane Perrio, Prof. Pierre Beslin and Dr Serge Masson for fruitful discussion on thiocarbonyl chemistry as well as students of our group at Caen for their efficient and enthusiastic contribution to this field.

References

1. Metzner P (1992) Synthesis 1185
2. Whitham GH (1995) Organosulfur chemistry. Oxford University Press, Oxford
3. Metzner P, Thuillier A (1994) Sulfur reagents in organic synthesis. Academic Press, London

4. Okazaki R (1995) In: Page PCB (ed) Organosulfur chemistry – synthetic aspects. Academic Press, London, p 225
5. Kirby GW (1993) Phosphorus, Sulfur Silicon Relat Elem 74:17
6. Kato S, Murai T (1992) In: Patai S (ed) The chemistry of acid derivatives – suppl B, vol 2, pt 1 (The chemistry of functional groups). Wiley, Chichester, p 803
7. Katritzky AR, Meth-Cohn O, Rees CW, Moody CJ (1995) Comprehensive organic functional group transformations. Pergamon, Oxford
8. Heathcock CH (1990) Aldrichim Acta 23:99
9. Schaumann E (1991) In: Trost BM, Fleming I, Winterfeldt E, Paquette LA (eds) Comprehensive organic synthesis, vol 6. Pergamon, Oxford, p 419
10. Voss J (1995) In: Paquette LA (ed) Encyclopedia of reagents for organic synthesis, vol 1. Wiley, Chichester, p 530
11. Voss J (1995) In: Paquette LA (ed) Encyclopedia of reagents for organic synthesis, vol 1. Wiley, Chichester, p 534
12. Whittingham WG (1995) In: Katritzky AR, Meth-Cohn O, Rees CW, Pattenden G (eds) Comprehensive organic functional group transformations, vol 3. Pergamon, Oxford, p 329
13. Ishii A, Nakayama J (1995) In: Katritzky AR, Meth-Cohn O, Rees CW, Moody CJ (eds) Comprehensive organic functional group transformations, vol 5. Pergamon, Oxford, p 505
14. Murai T, Kato S (1995) In: Katritzky AR, Meth-Cohn O, Rees CW, Moody CJ (eds) Comprehensive organic functional group transformations, vol 5. Pergamon, Oxford, p 545
15. Dell C (1995) In: Katritzky AR, Meth-Cohn O, Rees CW, Moody CJ (eds) Comprehensive organic functional group transformations, vol 5. Pergamon, Oxford, p 565
16. Owton WM (1995) In: Katritzky AR, Meth-Cohn O, Rees CW, Moody CJ (eds) Comprehensive organic functional group transformations, vol 5. Pergamon, Oxford, p 629
17. Vallée Y (1993) Rev Heteroatom Chem 8:1
18. McGregor WM, Sherrington DC (1993) Chem Soc Rev 199
19. Garcia Fernandez JM, Ortiz Mellet C (1996) Sulfur Rep 19:61
20. Hoeg-Jensen T (1996) Phosphorus, Sulfur Silicon Relat Elem 108:257
21. Kalaus G, Greiner I, Kajtar-Peredy M, Brlik J, Szabo L, Szantay C (1993) J Org Chem 58:6076
22. Ortiz Mellet C, Moreno Marin A, Garcia Fernandez JM, Fuentes J (1994) Tetrahedron: Asymmetry 5:2313
23. Michael JP, de Koning CB, Stanbury TV (1996) Tetrahedron Lett 37:9403
24. Hartke K, Gerber H-D (1996) J Prakt Chem 338:763
25. Bucher CB, Heimgartner H (1996) Helv Chim Acta 79:1903
26. Nieschalk J, Schaumann E (1996) Liebigs Ann Chem 141
27. Sakamoto M, Takahashi M, Kamiya K, Yamaguchi K, Fujita T, Watanabe S (1996) J Am Chem Soc 118:10,664
28. Sone H, Kondo T, Kiryu M, Ishiwata H, Ojika M, Yamada K (1995) J Org Chem 60:4774
29. Sharp MJ, Heathcock CH (1994) Tetrahedron Lett 35:3651
30. Seebach D, Ko SY, Kessler H, Köck M, Reggelin M, Schmieder P, Walkinshaw MD, Bölsterli JJ (1991) Helv Chim Acta 74:1953
31. Eberle MK, Jutzi-Eme A-M, Nuninger F (1994) J Org Chem 59:7249
32. Smith DC, Lee SW, Fuchs PL (1994) J Org Chem 59:348
33. Borthakur N, Goswani A (1995) Tetrahedron Lett 36:6745
34. Brillon D (1992) Synth Commun 22:1397
35. Le Roy-Gourvennec S, Masson S (1995) Synthesis 1393
36. Masson S (1994) Phosphorus, Sulfur Silicon Relat Elem 95/96:127
37. Barton DHR, Swift KAD, Tachdjian C (1995) Tetrahedron 51:1887
38. Makomo H, Masson S, Saquet M (1993) Phosphorus, Sulfur Silicon Relat Elem 80:31
39. Aloup J-C, Farge D, James C, Mondot S, Cavero I (1990) Drugs of the Future 15:1098
40. Hart TW, Guillochon D, Perrier G, Sharp BW, Vacher B (1992) Tetrahedron Lett 33:5117

41. Hart TW, Guillochon D, Perrier G, Sharp BW, Toft MP, Vacher B, Walsh RJA (1992) Tetrahedron Lett 33:7211
42. Katritzky AR, Drewniak M (1988) Tetrahedron Lett 29:1755
43. Katritzky AR, Moutou J-L, Yang Z (1995) Synthesis 1497
44. Katritzky A, Moutou J-L, Yang Z (1995) Synlett 99
45. Gronowitz S, Hörnfeldt A-B, Temciuc M (1993) Synthesis 483
46. Badudri F, Fiandanese V, Marchese G, Punzi A (1994) Synlett 719
47. Hartke K, Gerber H-D, Roesrath U (1991) Liebigs Ann Chem 903
48. Hartke K, Barrmeyer S (1996) J Prakt Chem 338:251
49. Le H-T, Mayer M, Thoret S, Michelot R (1995) Int J Peptide Protein Res 45:138
50. Hoeg-Jensen T, Olsen CE, Holm A (1994) J Org Chem 59:1257
51. Hoeg-Jensen T, Holm A, Sorensen H (1996) Synthesis 383
52. Shalaby MA, Grote CW, Rapoport H (1996) J Org Chem 61:9045
53. Hartke K (1991) Phosphorus, Sulfur Silicon Relat Elem 58:223
54. Voss J (1991) In: Trost BM, Fleming I, Winterfeldt E, Paquette LA (eds) Comprehensive organic synthesis, vol 6. Pergamon, Oxford, p 435
55. Kato S, Ishida M (1988) Sulfur Rep 8:155
56. Ramadas SR, Srinasavan PS, Ramachandran J, Sastry VVSK (1983) Synthesis 605
57. Kohrt A, Hartke K (1992) Liebigs Ann Chem 595
58. Hartke K, Brutsche A (1992) Synthesis 1199
59. Nakamura K, Kondo H, Metzner P, Ohno A (1992) Bull Inst Chem Res, Kyoto Univ 70:302
60. Villemin D, Thibault-Starzyk F (1993) Synlett 148
61. Moussounga JE, Bouquant J, Chuche J (1995) Bull Soc Chim Fr 132:249
62. Ennis DS, Armitage MA (1995) Tetrahedron Lett 36:771
63. Vallée Y, Khalid M, Ripoll J-L (1993) Tetrahedron Lett 34:2603
64. Guigné A, Metzner P (1990) Bull Soc Chim Fr 127:446
65. Pfister-Guillouzo G, Senio A, Gracian F, Khalid M, Ripoll J-L, Vallée Y (1995) New J Chem 19:1071
66. Hürzeler M, Bernet B, Vasella A (1993) Helv Chim Acta 76:995
67. Hürzeler M, Bernet B, Mäder T, Vasella A (1993) Helv Chim Acta 76:1779
68. Nicolaou KC (1996) Angew Chem, Int Ed Engl 35:589
69. Brunner A, Kühnle FNM, Seebach D (1996) Helv Chim Acta 79:319
70. Brutsche A, Hartke K (1992) Liebigs Ann Chem 921
71. Mouussounga J, Bouquant J, Chuche J (1994) Synthesis 483
72. Purkayastha ML, Chandrasekharam M, Vishwakarma JN, Hiriyakkanavar I, Junjappa H (1993) Synthesis 245
73. Bachi MD, Bosch E, Denemark D, Girsh D (1992) J Org Chem 57:6803
74. Capperucci A, Degl'Innocenti A, Ricci A, Mordini A, Reginato G (1991) J Org Chem 56:7323
75. Degl'Innocenti A, Capperucci A, Mordini A, Reginato G, Ricci A, Cerreta F (1993) Tetrahedron Lett 34:873
76. Sliwka H-R, Liaaen-Jensen S (1994) Acta Chem Scand 48:679
77. Le Nocher A-M, Metzner P (1992) Tetrahedron Lett 33:6151
78. Selzer T, Rappoport Z (1996) J Org Chem 61:5462
79. Bonini BF (1993) Phosphorus, Sulfur Silicon Relat Elem 74:31
80. Bonini BF, Busi F, de Laet RC, Mazzanti G, Thuring J-WJF, Zani P, Zwanenburg B (1993) J Chem Soc, Perkin Trans 1 1011
81. Bonini BF, Comes-Franchini M, Mazzanti G, Ricci A, Rosa-Fauzza L, Zani P (1994) Tetrahedron Lett 35:9227
82. Ishii A, Hoshino M, Nakayama J (1996) Pure Appl Chem 68:869
83. Ishii A, Akazawa T, Ding Mx, Honjo T, Maruta T, Nakamura S-y, Nagaya H, Ogura M, Teramoto K, Shiro M, Hoshino M, Nakayama J (1997) Bull Chem Soc Jpn 70:509
84. Ishii A, Akazawa T, Maruta T, Nakayama J, Hoshino M, Shiro M (1994) Angew Chem, Int Ed Engl 33:777
85. Ishii A, Jin Y-N, Nagaya H, Hoshino M, Nakayama J (1995) Tetrahedron Lett 36:1687

86. Ishii A, Akazawa T, Ding M-X, Honjo T, Nakayama J, Hoshino M (1993) J Am Chem Soc 115:4914
87. Capozzi G, Menichetti S, Nativi C, Rosi A, Franck RW (1993) Tetrahedron Lett 34:4253
88. Steliou K, Salama P, Yu X (1992) J Am Chem Soc 114:1456
89. Steliou K (1991) Acc Chem Res 24:341
90. Choi KS, Akiyama I, Hoshino M, Nakayama J (1993) Bull Chem Soc Jpn 66:623
91. Block E (1992) Angew Chem, Int Ed Engl 31:1135
92. Block E, Thiruvazhi M, Toscano PJ, Bayer T, Grisoni S, Zhao S-H (1996) J Am Chem Soc 118:2790
93. Block E, Bayer T, Naganathan S, Zhao S-H (1996) J Am Chem Soc 118:2799
94. Block E, Gillies JZ, Gillies CW, Bazzi AA, Putman D, Revelle LK, Wang D, Zhang X (1996) J Am Chem Soc 118:7492
95. Ferary S, Auger J (1996) J Chromatogr 750:63
96. Baudin J-B, Commenil M-G, Julia S, Wang Y (1992) Synlett 909
97. Baudin J-B, Commenil M-G, Julia S, Toupet L, Wang Y (1993) Synlett 839
98. Baudin J-B, Commenil M-G, Julia S, Wang Y (1996) Bull Soc Chim Fr 133:515
99. Braverman S, Grinstein D, Gottlieb HE (1994) Tetrahedron Lett 35:953
100. Pelloux-Léon N, Arnaud R, Ripoll J-L, Beslin P, Vallée Y (1997) Tetrahedron Lett 38:1385
101. Metzner P (1991) Phosphorus, Sulfur Silicon Relat Elem 59:1
102. Zwanenburg B (1989) Phosphorus, Sulfur Silicon Relat Elem 43:1
103. Le Nocher A-M, Metzner P (1991) Tetrahedron Lett 32:747
104. Cerreta F, Le Nocher A-M, Leriverend C, Metzner P, Pham TN (1995) Bull Soc Chim Fr 132:67
105. van der Linden JB, Timmermans JL, Zwanenburg B (1995) Recl Trav Chim Pays-Bas 114:91
106. Leriverend C, Metzner P (1994) Tetrahedron Lett 35:5229
107. Shi J, Tromm P-C, Heimgartner H (1994) Helv Chim Acta 77:2133
108. Lemarié M, Metzner P, Pham TN (1991) Tetrahedron Lett 32:7411
109. Bonini BF, Beulen MWJ, Comes-Franchini M, Mazzanti G, Van de Voort HHJM, Zwanenburg B (1996) Phosphorus, Sulfur Silicon Relat Elem 108:289
110. Chevrie D, Marrière E, Metzner P (1997) J Chem Soc, Perkin Trans 1 2019
111. Leriverend C, Metzner P, Capperucci A, Degl'Innocenti A (1997) Tetrahedron 53:1323
112. Nakayama J, Mizumura A, Yokomori Y, Krebs A, Schütz K (1995) Tetrahedron Lett 36:8583
113. Machiguchi T, Hasegawa T, Otani H (1994) J Am Chem Soc 116:407
114. Takeda N, Tokitoh N, Okazaki R (1997) Chem Eur J 3:62
115. Hasserodt J, Pritzkow H, Sundermeyer W (1993) Chem Ber 126:1701
116. Saito T, Nagashima M, Karakasa T, Motoki S (1992) J Chem Soc, Chem Commun 411
117. Okuma K, Shimasaki M, Kojima K, Ohta H, Okazaki R (1993) Chem Lett 1599
118. Huisgen R, Rapp J (1997) Tetrahedron 53:939
119. Anh NT (1995) Orbitales frontières - Manuel pratique. InterEditions CNRS Editions, Paris
120. Chiang Y, Jones J, Kresge AJ (1994) Tetrahedron Lett 116:8358
121. Tchertchian S, Vallée Y (1995) Tetrahedron Lett 36:6225
122. Murguia MC, Rossi RA (1997) Tetrahedron Lett 38:1355
123. Eames J, Jones RVH, Warren S (1996) Tetrahedron Lett 37:707
124. Beslin P, Perrio S (1993) Tetrahedron 49:3131
125. Beslin P, Perrio S (1992) Tetrahedron 48:4135
126. Walkup RD, Boatman PD (1993) Tetrahedron Lett 34:2417
127. Magnus P, Mendoza J (1992) Tetrahedron Lett 33:899
128. Metzner P (1989) Rev Heteroatom Chem 2:152
129. Berrada S, Metzner P, Rakotonirina R (1985) Bull Soc Chim Fr 881
130. Oare DA, Henderson MA, Sanner MA, Heathcock CH (1990) J Org Chem 55:132
131. Oare DA, Heathcock CH (1991) Top Stereochem 20:87
132. Viteva LZ, Gospodova TS, Stefanovsky YN (1994) Tetrahedron 50:7193
133. Heathcock CH, Davidsen SK, Mills SG, Sanner MA (1992) J Org Chem 57:2531

134. Mook RA, Lackey K, Bennett C (1995) Tetrahedron Lett 36:3969
135. Maggini M, Prato M, Ranelli M, Scorrano G (1992) Tetrahedron Lett 33:6537
136. Beslin P, Marion P (1992) Tetrahedron Lett 33:935
137. Brutsche A, Hartke K, Gerber H-D (1992) Liebigs Ann Chem 927
138. Beslin P, Marion P (1992) Tetrahedron Lett 33:5339
139. Hartke K, Bien N, Weller F (1996) Z Naturforsch, Teil B 51:1173
140. Mazzanti G, van Helvoirt E, Van Vliet LA, Ruinaard R, Masiero S, Bonini BF, Zwanenburg B (1994) J Chem Soc, Perkin Trans 1 3299
141. Creary X, Zhu C (1995) J Am Chem Soc 117:5859
142. Beak P, Park YS, Liu C (1994) J Org Chem 59:7410
143. Shi J, Heimgartner H (1996) Helv Chim Acta 79:371
144. Portella C, Shermolovich Y (1997) Tetrahedron Lett 38:4063
145. Capperucci A, Ferrara M-C, Degl'Innocenti A, Bonini B, Mazzanti G, Zani P, Ricci A (1992) Synlett 880
146. Capperucci A, Ferrara M-C, Degl'Innocenti A, Bonini B, Mazzanti G, Zani P, Ricci A (1994) Tetrahedron Lett 35:161
147. Hagen JP (1993) J Org Chem 58:506
148. Calderwood DJ, Davies RV, Rafferty P, Twigger HL, Whelan HM (1997) Tetrahedron Lett 38:1241
149. Metzner P (1996) Pure Appl Chem 68:863
150. Alayrac C, Cerreta F, Corbin F, Chapron I, Metzner P (1996) Tetrahedron Lett 37:4507
151. Capperucci A, Degl'Innocenti A, Leriverend C, Metzner P (1996) J Org Chem 61:7174
152. Wagner K, Weiß D, Beckert R (1995) Synthesis 1245
153. Delêtre M, Levesque G (1990) Macromolecules 23:4733
154. Papandreou G, Tong MK, Ganem B (1993) J Am Chem Soc 115:11,682
155. Masson S (1994) Main Group Chem News 2(2):18
156. Makomo H, Masson S, Saquet M (1994) Tetrahedron 50:10,277
157. Makomo H, Masson S, Saquet M (1993) Tetrahedron Lett 34:7257
158. Wilker S, Laurent C, Sarter C, Puke C, Erker G (1995) J Am Chem Soc 117:7293
159. Nielsen JK, Madsen JO (1994) Tetrahedron:Asymmetry 5:403
160. Duchenet V, Vallée Y (1993) J Chem Soc, Chem Commun 806
161. Cerreta F, Leriverend C, Metzner P (1993) Tetrahedron Lett 34:6741
162. Baltas M, Raouf-Benchekroun K, De Blic A, Cazaux L, Tisnès P, Gorrichon L, Hussein K, Barthelat J-C (1996) Tetrahedron 52:14,865
163. Baudin J-B, Commenil M-G, Julia SA, Lorne R, Ruel O (1996) Bull Soc Chim Fr 133:1127
164. Makioka Y, Uebori S-y, Tsuno M, Taniguchi Y, Takaki K, Fujiwara Y (1996) J Org Chem 61:371
165. Makioka Y, Uebori S-y, Tsuno M, Taniguchi Y, Takaki K, Fujiwara Y (1994) Chem Lett 611
166. Makioka Y, Tsuno M, Takaki K, Taniguchi Y, Fujiwara Y (1995) Chem Lett 821
167. Padwa A, Weingarten MD (1996) Chem Rev 96:223
168. Kim G, Kang S, Keum G (1994) Tetrahedron Lett 35:3747
169. Padwa A, Kinder FR, Zhi L (1991) Synlett 287
170. Nkiliza J, Vercauteren J, Léger J-M (1991) Tetrahedron Lett 32:1787
171. Artis DR, Cho I-S, Jaime-Figueroa S, Muchowski JM (1994) J Org Chem 59:2456
172. Ravindranathan T, Chavan SP, Awachat MM, Kelkar SV (1995) Tetrahedron Lett 36:2277
173. Bonnat M, Durand J-O, Le Corre M (1996) Tetrahedron: Asymmetry 7:559
174. Kuroboshi M, Hiyama T (1992) Chem Lett 827
175. Kuroboshi M, Hiyama T (1994) Tetrahedron Lett 35:3983
176. Kim K-I, McCarthy R (1996) Tetrahedron Lett 37:3223
177. Furuta S, Hiyama T (1996) Synlett 1199
178. Rozen S, Mishani E (1993) J Chem Soc, Chem Commun 1761
179. Kuroboshi M, Hiyama T (1994) Synlett 251
180. Shiosaki K (1991) In: Trost BM, Fleming I, Paquette LA, Heathcock CH (eds) Comprehensive organic synthesis, vol 2. Pergamon, Oxford, p 865
181. Munchhof MJ, Meyers A (1995) J Am Chem Soc 117:5399

182. Marchand P, Fargeau-Bellasoued M-C, Bellec C, Lhommet G (1994) Synthesis 1118
183. Smith DC, Fuchs PL (1995) J Org Chem 60:2692
184. Michael JP, Gravestock D (1996) Synlett 981
185. Takahata H, Suzuki T, Yamazaki T (1986) Heterocycles 24:1247
186. Sosnicki JG, Liebscher J (1996) Synlett 1117
187. Hart DJ, Sun L-Q, Kozikowski AP (1995) Tetrahedron Lett 36:7787
188. Motherwell WB, Crich D (1992) Free radical chain reactions in organic synthesis. Academic Press, London
189. Crich D, Quintero L (1989) Chem Rev 89:1413
190. Zard SZ (1997) Angew Chem, Int Ed Engl 36:672
191. Nicolaou KC, Sato M, Theodorakis EA, Miller ND (1995) J Chem Soc, Chem Commun 1583
192. Bachi MD, Melman A (1997) J Org Chem 62:1896
193. Feldman KS, Schildknegt K (1994) J Org Chem 59:1129
194. Yamabe S, Okumoto S, Hayashi T (1996) J Org Chem 61:6218
195. Arnaud R, Dillet V, Pelloux-Léon N, Vallée Y (1996) J Chem Soc, Perkin Trans 2 2065
196. Devine PN, Meyers AI (1994) J Am Chem Soc 116:2633
197. Reddy KV, Rajappa S (1992) Tetrahedron Lett 33:7957
198. Désert S, Metzner P, Ramdani M (1992) Tetrahedron 48:10,315
199. Désert S, Metzner P (1992) Tetrahedron 48:10,327
200. Beslin P, Perrio S (1989) J Chem Soc, Chem Commun 414
201. Beslin P, Perrio S (1991) Tetrahedron 47:6275
202. Alayrac C, Fromont C, Metzner P, Anh NT (1997) Angew Chem, Int Ed Engl 36:371
203. Alayrac C, Nowaczyk S, Lemarié M, Metzner P (1999) Synthesis 669
204. Hwu JR, Anderson DA (1991) J Chem Soc, Perkin Trans 1 3199
205. Schatz J, Sauer J (1994) Tetrahedron Lett 35:4767
206. Wilker S, Erker G (1995) J Am Chem Soc 117:10,922
207. Bonini BF, Masiero S, Mazzanti G, Zani P (1991) Tetrahedron Lett 32:2971
208. Degl'Innocenti A, Scafato P, Capperucci A, Bartoletti L, Spezzacatena C, Ruzziconi R (1997) Synlett 361
209. Arnaud R, Chavant PY, Molvinger K, Vallée Y (1995) J Chem Soc, Chem Commun 1897
210. Barone V, Arnaud R, Chavant PY, Vallée Y (1996) J Org Chem 61:5121
211. Tamaru Y, Sakata H, Kimura M, Harayama H, Konishi H, Fugami K, Tanaka S (1994) J Chem Soc, Chem Commun 2365
212. Tamaru Y, Harayama H, Sakata H, Konishi H, Fugami K, Kimura M, Tanaka S, Okajima T, Fukazawa Y (1997) Liebigs Ann Recueil 907
213. Kirby G, Mahajan MP, Rahman MS (1991) J Chem Soc, Perkin Trans 1 2033
214. Kirby GW, Lochead AW, Willamson S (1996) J Chem Soc, Perkin Trans 1 977
215. Morita H, Takeda M, Kamiyama H, Hashimoto T, Yoshimura T, Shimasaki C, Tsukurimichi E (1996) Tetrahedron Lett 37:3739
216. Sato R, Satoh S-i (1991) Synthesis 785
217. Roxburgh CJ (1996) Synthesis 307
218. Pinto IL, Buckle DR, Rami HK, Smith DG (1992) Tetrahedron Lett 33:7597
219. Takahashi T, Kurose N, Koizumi T (1993) Heterocycles 36:1601
220. Larsen SD, Fisher PV, Libby BE, Jensen RM, Mizsak SA, Watt W (1996) J Org Chem 61:4725
221. Abelman MM (1991) Tetrahedron Lett 32:7389
222. Saito T, Shizuta T, Kikuchi H, Nakagawa J, Hirotsu K, Ohmura H, Motoki S (1994) Synthesis 727
223. Saito T, Karakasa T, Fujii H, Furuno E, Suda H, Kobayashi K (1994) J Chem Soc, Perkin Trans 1 1359
224. Motoki S, Saito T, Karakasa T, Kato H, Matsushita T, Hayashibe S (1991) J Chem Soc, Perkin Trans 1 2281
225. Saito T, Fujii H, Hayashibe S, Matsushita T, Kato H, Kobayashi K (1996) J Chem Soc, Perkin Trans 1 1897
226. Saito T, Kawamura M, Nishimura J-i (1997) Tetrahedron Lett 38:3231

227. Saito T, Kimura H, Sakamaki K, Karasaka T, Moriyama S (1996) Chem Commun 811
228. Rapp J, Huisgen R (1997) Tetrahedron 53:961
229. Nakayama J, Akimoto K (1994) Sulfur Rep 16:61
230. Okuma K, Yamamoto T, Shirokawa T, Kitamura T, Fujiwara Y (1996) Tetrahedron Lett 37:8883
231. Bell AS, Fishwick CWG, Reed JE (1996) Tetrahedron Lett 37:123
232. Marchand A, Mauger D, Guingant A, Pradère J-P (1995) Tetrahedron: Asymmetry 6:853
233. Marchand A, Pradère J-P, Guingant A (1997) Tetrahedron Lett 38:1033
234. Thompson AM, Boyd M, Denny WA (1993) J Chem Soc, Perkin Trans 1 1835
235. Szantay C, Moldvai I, Fishwick CWG, Hill DR (1991) Tetrahedron Lett 32:2529
236. Ohno M, Kojima S, Eguchi S (1995) J Chem Soc, Chem Commun 565
237. Capozzi G, Franck RW, Mattioli M, Menichetti S, Nativi C, Valle G (1995) J Org Chem 60:6416
238. Capozzi G, Falciani C, Menichetti S, Nativi C, Franck RW (1995) Tetrahedron Lett 36:6755
239. Capozzi G, Falciani C, Menichetti S, Nativi C (1997) J Org Chem 62:2611
240. Capozzi G, Dios A, Franck RW, Geer A, Marzabadi C, Menichetti S, Nativi C, Tamarez M (1996) Angew Chem, Int Ed Engl 37:777
241. Barbaro G, Battaglia A, Giorgianni P, Bonini BF, Maccagnani G, Zani P (1991) J Org Chem 56:2512
242. Schmid T, Hanack M, Maichle C, Strähle J (1993) Angew Chem, Int Ed Engl 32:258
243. Capozzi G, Fratini P, Menichetti S, Nativi C (1995) Tetrahedron Lett 36:5089
244. Capozzi G, Fratini P, Menichetti S, Nativi C (1996) Tetrahedron 57:12,233
245. Capozzi G, Fratini P, Menichetti S, Nativi C (1996) Tetrahedron 57:12,247
246. Choi SS-M, Kirby GW (1991) J Chem Soc, Perkin Trans 1 3225
247. Choi SS-M, Kirby GW, Mahajan M (1992) J Chem Soc, Perkin Trans 1 191
248. Saito T, Watanabe T, Kitazawa S, Hayashi Y, Motoki S (1991) J Chem Soc, Perkin Trans 1 959
249. Pattenden G, Shuker AJ (1991) Synlett 717
250. Fisera L, Huisgen R, Kalwinsch I, Langhals E, Li X, Mloston G, Polborn K, Rapp J, Sicking W, Sustmann R (1996) Pure Appl Chem 68:789
251. Mloston G, Heimgartner H (1995) Helv Chim Acta 78:1298
252. Mloston G, Romanski J, Linden A, Heimgartner H (1995) Helv Chim Acta 78:1067
253. Mloston G, Romanski J, Linden A, Heimgartner H (1995) Helv Chim Acta 78:1499
254. Mloston G, Romanski J, Linden A, Heimgartner H (1993) Helv Chim Acta 76:2147
255. Mloston G, Romanski J, Linden A, Heimgartner H (1996) Helv Chim Acta 79:1305
256. Huisgen R, Fisera L, Giera H, Sustmann R (1995) J Am Chem Soc 117:9671
257. Sustmann R, Sicking W, Huisgen R (1995) J Am Chem Soc 117:9679
258. Tabuchi T, Nojima M, Kusabayashi S (1991) J Chem Soc, Perkin Trans 1 3043
259. Mloston G, Linden A, Heimgartner H (1996) Helv Chim Acta 79:31
260. Huisgen R, Mloston G, Polborn K (1996) J Org Chem 61:6570
261. Huisgen R, Mloston G, Polborn K, Sustmann R, Sicking W (1997) Liebigs Ann Recueil 179
262. Huisgen R, Mloston G, Polborn K, Palacios-Gambra F (1997) Liebigs Ann Recueil 187

Author Index Volume 201–204

Author Index Vols. 26–50 see Vol. 50
Author Index Vols. 51–100 see Vol. 100
Author Index Vols. 101–150 see Vol. 150
Author Index Vols. 151–200 see Vol. 200

The volume numbers are printed in italics

MIX
Papier aus verantwortungsvollen Quellen
Paper from responsible sources
FSC® C105338

If you have any concerns about our products,
you can contact us on
ProductSafety@springernature.com

In case Publisher is established outside the EU,
the EU authorized representative is:
Springer Nature Customer Service Center GmbH
Europaplatz 3, 69115 Heidelberg, Germany

Printed by Libri Plureos GmbH
in Hamburg, Germany